THE
METALCASTER'S
BIBLE

Other TAB books by the author:

No. 725   *Lost Wax Investment Casting*
No. 1043   *The Complete Handbook of Sand Casting*

Dedication

This book is dedicated to Hazel, my wife, who contributes very, very much to my writings.

Acknowledgements

The author wishes to thank the following companies for technical information and drawings supplied for use in this manuscript.

NL Baroid/NL Industries, Inc.
Royer Foundry & Machine Co.
Wheelabrator-Frye Inc.
Joseph Dixon Co.

No. 1173
$15.95

THE METALCASTER'S BIBLE

BY C. W. AMMEN

TAB BOOKS

BLUE RIDGE SUMMIT, PA. 17214

FIRST EDITION

FIRST PRINTING—MARCH 1980

Copyright © 1980 by TAB BOOKS

Printed in the United States of America

Library of Congress Cataloging in Publication Data

Ammen, C. W.
　　The metalcaster's bible.

　　Includes index.
　　1. Founding.　　I. Title.
TS230.A63　　　　671.2　　　　79.25217
ISBN 0-8306-9970-8
ISBN 0-8306-1173-8 pbk.

Contents

Introduction

If you are a fledging or novice metal caster and wish to cast for your own amazement, the hobby foundry should be your cup of tea. The hobby foundry can be fully or partially self sufficient. You can cast items which you sell as your product, like a sculptor who casts his own work, or you can cast for the fun of it! There are numerous small backyard foundries today in this country. For instance, if you need a small casting for a product you make for someone, or yourself, but the quantity is too small to interest a production shop, or job shop, where do you turn?

Not too many years ago, thousands of independent foundries flourished in the United States. Many were jobbing foundries: the foundry cast one or several items as an assignment from a customer. Jobbers produced castings weighing anywhere from a few ounces to many tons. The customer could specify brass, bronze, aluminum, iron or steel.

In time, jobbers became more specialized. Some cast only ferrous metals, others nonferrous. They also became selective as to the types and weights of the castings they would produce. As industry grew in size and complexity, it could no longer deal with the jobber who made "one of this and 10 of that." It needed hundreds and thousands of one casting. Soon it became necessary to build and operate company foundries so that specialized needs could be met, large quantities guaranteed and cost-control insured. The *captive shop* foundry was born.

Foundries, however, are on the decline today. Of the thousands in the country a decade ago, only a fraction are left. But new casting techniques and mechanization has increased the tonnage-per-year produced. The jobbing foundry will be extinct in the not too distant future. This is due to, among other factors, the state of the economy, the loss of skilled molders (who could cast anything from a doorknob to a 6-ton gear), and the lack of training for fledgling founders.

There is still a demand for one-shot jobbing, but these jobs go begging. Large foundries (captive or independent) cannot engage in this sort of operation; not only do they lack the know-how for single-piece production, they cannot do it profitably. The skilled sand and investment molder has been replaced by a machine operator who is hardly aware of what the melting and casting art is all about.

So great is the void to be filled that interest in all types of do-it-yourself casting is fast reaching an all-time high.

If you are interested in casts of items for personal use, it would be foolish of you to have a foundry produce it. Due to the price of small quantities, foundry prices would more than likely be more than you could stand, especially compared to the retail price for the unit.

But if your requirement for parts is extensive, or if you wish to explore this fascinating work as a creative hobby, you are encouraged to undertake the work on your own.

Casting metal parts can be one of the most creative and rewarding activities that the home craftsman can undertake. Few amateurs have attempted this work. They probably assume that the operations require too much equipment, skill and know-how. Fortunately, however, foundry work is not beyond the means or capabilities of the home craftsman, and there is no valid reason why these useful arts cannot find its way into your home shop.

As this book is not intended for the professional or the person who declares he is going to learn to metal cast even if it takes a week, my advice is plain and simple. Simply start out slowly. You really don't have to know exactly what your ultimate goal is, if any.

It is wise to warn you right now that foundry operations of the kind we will be discussing in the chapters that follow is often hard, dirty work, but anyone with average strength and manual dexterity should be able to handle foundry work with ease.

C. W. Ammen

Part 1
The Process of Metal Casting

Chapter 1
Casting Processes

Of the total tonnage of castings produced each year, the greatest percentage is produced by sand casting.

The Chinese produced a large quantity of extremely fine and complicated bronzes, some of which were quite massive. Careful, unbiased studies of these bronzes through X-ray examination and chemical analysis of the core materials has been most enlightening. Some consist of many sand castings assembled with cleverly designed hidden joints (undetectable except by X-ray examination). The surface texture, in the majority of cases, is not an as-cast finish, but a carefully chased and hand-engraved one. The mold material consisted of fine silica and natural cement (a sand loam) and in some cases, a natural bonded green sand very similar to French sand. Not only where the Chinese clever at making joints, they made some highly complicated piece molds.

Remember when examining a bronze, that you are not looking at the item as cast but, rather, the finished chased and patined work: it might have been produced by any number of methods. The T. F. McGann & Sons Co. of Boston produced countless fine statuary bronzes in French molding sand by the green sand molding process.

New York is the home of over 300 examples of buildings made from iron members cast in sand; some highly ornamented, were produced before structural steel was widely used. The French Quarter of New Orleans has fine examples of balconies and fences commonly thought to be wrought iron; they are for the most part not wrought iron at all, but gray iron cast in sand. Some were produced in France from finely carved wooden patterns.

GREEN SAND CASTING

Green sand molds are made in two-part flasks. The pattern is placed on a smooth molding board and covered with an inverted drag flask; green sand is tamped down firmly into drag flask, leveled off with the bottom, and covered with a bottom board. The boards and flask are then overturned and the molding board removed.

This leaves the upper face of the pattern at the flask parting plane. The cope is placed in register on the drag and the sand face of the drag flask is given a light coating of parting powder (a commercial product) to keep the two halves of the mold from sticking together.

A sprue stick is placed in position with the cope and surrounded with sand tamped firmly to the top of the cope flask. After the sand is leveled off, the sprue stick is removed. The cope flask is lifted off and set aside; the pattern is removed from the drag flask and a gate is cut over to the sprue for the entrance of metal to the mold cavity. The entire mold is blown clean of any loose sand and the cope is replaced. The mold is then ready to pour. Weights are placed on the cope to prevent it from lifting when poured (and causing a *run out*).

After pouring, the metal is allowed to solidify, the mold is shaken out, and the casting and gate are removed.

SAND CASTING

Crude methods of sand casting were practiced before the beginning of recorded history.

Sand casting consists of pouring a molten metal or alloy into a mold of earth or sand and allowing it to solidify. Molding material is cheap and plentiful. It is also readily worked into suitable molds, making sand casting the most economical method if only one or a few parts are to be made. Machining or building up a part by welding, riveting or brazing is surely more expensive than casting. Sand casting is also an extremely versatile process. Parts can be made in almost any size varying from a fraction of an ounce up to tons in weight.

Due to the lack of grain coherence, pure sand would not be suitable for making molds. A binder is needed to hold the sand particles together. Clay is perfectly suited for this purpose. However, an abundance of clay is undesirable because the sand then loses its porosity and porosity is essential for the escape of gases when the molten metal is poured into the mold.

Sand Casting (Chemical Binder)

Chemically bonded molds are produced by mixing a chemical binder with clean dry silica sand and adding a catalyst to harden the binder.

Most widely used today are the furan binders mixed with an 85 percent phosphoric acid as a catalyst.

It is a very good medium for large bronzes and can be used with oil and clay patterns, as well as wood or plastic patterns. It has little strength in the green state and cannot be used like green sand. You must make up the drag half of the mold, allow it to set and then remove the cope and finish the mold as usual. It has no equal for piece molds and highly complicated molds, once you get the hang of it.

Its high strength when cured allows relatively thin molds for heavy section castings. Because no moisture is involved, the yield of good castings is extremely high. The basic formula is:

Binder—2 percent of the weight of sand
Catalyst—25 percent of the weight of the binder

The binder is furfural alcohol: the catalyst, an 85 percent phosphoric acid solution.

First mix the sand with the acid, then add the binder and mix thoroughly. You have 20 or 30 minutes to work before the sand sets. The catalyst reacts with the binder to produce a hard resin, which cements the grains of sand together to form a hard, yet permeable mold. Any metal can be cast in furan molds.

Unlike green sand molds, where the sand can be retempered and used over again to produce new molds, furan molds can be used only once.

Sand Casting With Oil Binders

Molds and cores can be made by mixing silica sand with linseed oil or any number of commercial drying oils. As this is the usual method for producing cores, a mold made with a *drying* oil binder, such as linseed oil and silica sand, is called a *core mold*. Due to its low green strength, it requires special pattern equipment and rigging; it has been largely replaced with furan methods. To make such a mold mix 40 parts (by volume) of clean silica sand with one part (by volume) of linseed oil. Dry the mold completely at 350°F. The oil oxidizes during drying, binding the sand grains together. Like furan molds, it is a one-time material. Unlike furan, oil binders require baking to get a bond.

Sand Casting With Styrofoam

In this method, the object to be cast or reproduced in metal is first made as a Styrofoam pattern. The Styrofoam pattern is surrounded with green sand in a flask. When molten metal enters the mold, the heat vaporizes the Styrofoam.

This system is ideally suited to large and chunky castings due to its simplicity. However, its voluminous output of noxious smoke has caused considerable grumbling from the EPA; and, so I am told, is outlawed in some areas. Done on a small scale, and infrequently, it might be tolerated. You need a new pattern for each casting.

Sand Casting With Carbon Dioxide

The CO_2 system, or silicate of soda method, is like the furan system in that it is very widely used.

When CO_2 contacts the water in silicate of soda, carbonic acid is formed. The acid neutralizes the silicate of soda, forming a glass bond for the molding sand.

Cores and molds may require as little as 2½ percent silicate of soda binder, to as much as 4 percent, depending upon the mass involved. Various grades and formulations of silicate of soda are available for a wide range of applications. The advantage over furan binders is the short time it takes to "gas" a mold or core—in some cases, 10 seconds or less.

After the sand and silicate of soda are gassed, the mold is ready to pour. This makes for very rapid production. It is not as versatile as the furan method, in that manifolds, probes, a tank of CO_2, a regulator, hoses and gassing cups are required.

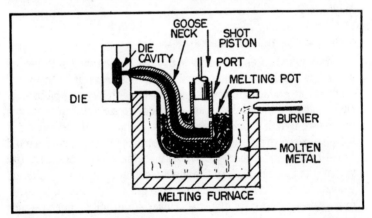

Fig. 1-1. Hot chamber die casting.

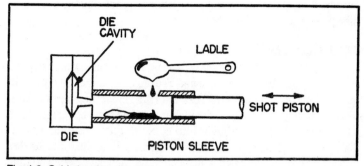

Fig. 1-2. Cold chamber die casting.

DIE CASTING

Die casting is a very rapid method of producing large quantities of castings. As the name implies, the mold cavity is a *die*—a permanent metal mold into which the metal is introduced under pressure.

The process is suitable only to relatively high production where a large quantity of a given casting is desired. The initial cost to make the die and the high cost of die-casting machines make it imperative that sufficient quantities of a given casting are needed to justify the high cost.

Basically, there are two distinct types of die-casting equipment: the *hot chamber* and the *cold chamber*. In the hot chamber (Fig. 1-1), the mechanism which shoots the metal into the die to produce each cast is submerged under the molten metal in the machine's melting pot. This makes it automatic in the sense that the "shot" cylinder is automatically filled with metal for each shot, making it unnecessary for the operator to ladle metal into the cylinder for each casting or shot.

The cold chamber machine (Fig. 1-2) requires that the shot cylinder be filled manually by the operator for each casting (or shot).

The required amount of metal is poured into the shot cylinder and the shot piston is advanced, shooting the metal into the die cavity.

PLASTER MOLD CASTING

Plaster mold casting has a number of applications. The method is a variation of sand casting. Instead of using sand to make the mold a mixture of plaster, talc and water flow up around the pattern. After the plaster has set, the pattern is withdrawn. The mold is then baked in an oven, driving off moisture. This process is more expensive than sand casting, but it does offer a higher degree of dimensional accuracy.

SHELL MOLD CASTING

Shell molding is a fairly new process in the foundry industry. It is sometimes called the Croning Process after the inventor. It makes use of a mixture of sand and thermosetting resin. This mixture is usually phenol-formaldehyde and it forms the mold. This sand-resin mixture is brought into contact with a heated pattern. The mixture closest to the pattern forms a thin shell due to the polymerization of the resin which binds the sand particles. The thin shell is then used as a mold, backed up by loose sand or shot to give it strength. Excellent finish and dimensional accuracy are thus obtained. Shell molding is a large-scale industrial process because of the required expensive production machinery.

SLUSH CASTING

In slush casting, a metal mold in two or more parts is used. The mold is filled with metal and rocked back and forth until the metal (touching the mold surfaces) solidifies in the desired thickness. When this point is reached, the liquid metal remaining in the mold is poured out, resulting in a hollow casting. This system is used only for low-melting lead and zinc-based metals, and to produce ornamental items that need not be strong.

This system is similar to the process used to slip cast clay pottery. The slip in this case is metal. The mold is made of metal rather than plaster of paris.

PERMANENT MOLDS

Permanent molds are metal molds designed to be opened and closed. The metal is poured into the mold, and when solidified, the mold is opened and the casting is removed; the cycle is then repeated.

Two very common types of permanent molds which have been around for many years are those used to make bullets and sinkers for fishing. However, they are not limited to these types of castings. Hundreds of thousands of different kinds of permanent mold castings in many sizes are made each year. Over 90 percent of all automotive aluminum pistons and most cast aluminum cookware are produced in cast-iron permanent molds.

There are other casting methods; but by and large, the methods described account for the largest portion of the castings produced today.

CASTING DEFECTS

A knowledge of casting defects is essential. If you cannot pinpoint the cause of a defect there is no way of correcting the problem. Some defects are quite obvious, along with the cause. Some types of defects can often resemble each other in appearance and separating them is often difficult. A drawing or photograph of a defect is one thing and looking at the actual defect is another. You learn about defects by analyzing your own and those of others.

Here is a list of defects and their causes that should include anything you might ever encounter:

- **Poured Short.** Casting incomplete due to not filling the mold. This is a stupid trick which we all do. It causes insufficient metal in the ladle. Going back and touching up the mold will not do it.

- **Bobble.** Here the casting was poured short. The problem was a slacked or interrupted pour. When pouring a casting the metal must be poured at a constant choked velocity. If you slack off or reduce the velocity, you can cause this defect. A completely interrupted pour, start—stop—start, if only for a second. A great percentage of lost castings is caused by pouring improperly.

- **Slag Inclusion.** Slag on the face of the casting and usually down the sides of the sprue. The cause is not skimming the ladle properly, not choking the sprue (keeping it brimming full from start to finish), sprue too big (cannot be kept choked) and gating system improperly choked.

- **Steam Gas Porosity.** Usually shows up as round holes like swiss cheese, just under the cope surface of the casting and comes to light during machining. The cause is a wet ladle, where the ladle lining was not properly and thoroughly dried.

  In extreme cases the metal will kick and boil in the ladle. The practice of pigging the metal in a wet ladle and refilling it with the hopes that the pigged metal finished drying the ladle is sheer folly.

- **Kish.** (cast iron) If the carbon equivalent of the iron is too high for the section poured and its cooling rate is slow, free graphite will form on the cope surface in black shiny flakes free from the casting, causing rough holey defects usually widespread. Carbon equivalent is the relationship of the total carbon to the silicon and phosphorous content of the iron, which is controlled by the make up charge and silicon added at the spout. It is called carbon equivalent because the addition of silicon or phosphorus is only one third as effective as carbon, therefore the carbon equivalent of the three additives is equal to the total percentage of pure carbon plus one third the percentage of the silicon and phosphorus combined. The carbon equilvalent is varied by the foundryman depending on what type of iron he is producing and what he is pouring.

- **Inverse Chill.** Found in gray iron castings and is hard or chilled iron in the center sandwiched between soft iron. The cause again could be incorrect carbon equivalent for the job or the presence of nonferrous metals in the charge, lead, antimony or tellurium which are detrimental impurities.

- **Broken Casting.** Could be caused by improper design. Improper filleting along with improper handling any where along the line. Copper base castings, red brass, yellow brass, etc., are what are known as hot short. They will break easily when hot. Thus, if the casting is shaken out of the mold before it has cooled sufficiently, it can get broken very easily. A mold or core which has a too high hot strength will not give or collapse to give the casting room to move as it shrinks and will break the casting.

- **Bleeder.** Caused by shaking the casting out too soon when a portion of it is still liquid. The section runs out leaving a defect. In extreme cases the entire center section will run out on the shake out man's feet.

- **Lead Sweat.** A covering of lead tears on the outside of a high leaded bronze or brass casting with underlying holes or porosity (red metals containing a large percentage of lead such as leaded bearing bronze). Since the lead has the lowest melting point of the constituents of high leaded bronze (copper 70 percent—tin 5 percent—lead 25 percent) and is not in solution with the copper and tin, it remains liquid until the casting has cooled below 620 degrees Fahrenheit. If the casting is shaken out before it is below this temperature the lead will sweat out from between the copper and tin crystals.

- **Run Out.** Caused by the metal in the mold running out

between the joint of the flask which drains the liquid metal partially or completely from any portion of the casting above the parting line. A run out can come through or between the drag and the bottom board from a cracked drag mold. Also, it can come from between a loose improperly fitted core and the core print. It is caused by insufficient room between the flask and the cavity, insufficient weight on the cope (cope raises during pouring) or improperly clamped molds. Excessive hydrostatic pressure (sprue too tall for the job), no dough roll used between cope and drag (large jobs) or a combination of all of the above.

An attempt to save a run out by placing your foot on top of the cope and applying pressure above the point where the liquid metal is running out is foolhardy and dangerous. Also trying to stop off the run out flow with sand or clay is folly.

- **Omission of a Core.** Results and cause obvious.
- **Ram Off.** A defect resulting from a section of the mold being forced away from the pattern by ramming sand after it has conformed to the pattern contour. This is caused by careless ramming where the mold is rammed vertically and then on an angle, causing the vertically rammed sand to slide sideways leaving a gap between the pattern and the sand, resulting in a deformed casting. Another cause is using a sand with poor or low flowability. (Too much clay or sand too fine.)
- **Core Rise.** Caused by a core rising from its intended position toward the cope surface, causing a variation in wall thickness or if when touching the cope, there is no metal at that point. The core has shifted from its position. A green sand core rise is when a green sand core in the drag is cracked at its base (caused when drawing the pattern) and floats toward the cope. Dry sand cores will float if the unsupported span of a thin insufficiently rodded core is too great. It will bend upward by the buoyancy of the metal. Insufficient core prints in number and design, insufficient chaplets, slipped chaplet, chaplets left out by molder, poor design of the core. This defect is easy to spot and remedy.
- **Shifts.** These come in two classifications: *mold shift* and *core shift*.

A mold shift is when the parting lines are not matched when the mold is closed, resulting in a casting offset or mismatched at the parting. The causes are: Excessive rapping of a loose pattern, reversing the cope on the drag, too loose a fit of the pattern pins and dowels, faulty mismatched flasks, too much play between pins and guides, faulty clamping, improper fitting (racked) jackets and improper placing of jackets.

A core shift is caused by not aligning the halves of glued cores true and proper when assembling them.

- **Swell.** In this defect you have a casting which is deformed due to the pressure of the metal moving or displacing the sand. It is usually caused by a soft spot or a too soft mold.

- **Sag.** A decrease in metal section due to a core or the cope sagging. The cause is insufficient cope bars, too small a flask for the job or insufficient cope depth. This defect will also cause misruns.

- **Fin.** A fin of metal on the casting caused by a crack in the cope or drag. It is caused by wreched flasks, bad jackets, (and setting) uneven warped bottom boards, uneven strike off, insufficient cope or drag depth, bottom board not properly rubbed or drag mold sitting uneven (rocking).

- **Fushion.** A rough glassy surface of fused melted sand on the casting surface either on the outside or on a cored surface. The cause is a too low sintering or melting point of the sand or core. This is quite common when a small diameter core runs through an exceptionally heavy section (of great heat) which actually melts the core. This can also be caused by pouring much too hot (hotter than necessary) for the sand or cores. A mold or core wash can prevent this in some cases but if the sintering point of your sand is too low for the class of work, you need a more refractory sand, zircon, etc.

- **Metal Penetration.** This defect should not be confused with an expansion scab which is attached to the casting by a thin vein of metal.

  The defect is a rough unsightly mixture of sand and metal caused by the metal penetrating into the mold wall or the surfaces of a core (not fused).

  The defect is basically caused by too soft and uneven ramming of the mold or core, making the sand too (open) porous, also a too high pouring temperature, a too sharp corner (insufficient filleting), making it impossible to ram the sand tight enough, localized overheating of the sand due to poor gating practice or molding with a sand too open for the job.

Penetration in brass castings is sometimes traced to excessive phosphorus used in de-oxidizing the metal, making it excessively fluid.

● **Rough Surface.** Can run from mild penetration to spotty rough spots or a completely rough casting. Many factors or combinations are at fault—sand too coarse for the weight and pouring temperature of the casting, improperly applied or insufficient mold coating or core coating, faulty finishing, excessive use of parting compound (dust), hand cut gates not firm or cleaned out, dirty pattern, sand not riddled when necessary, excessive or too coarse of sea coal in the sand, permeability too high for class of work or core or mold wash faulty (poor composition).

● **Blows.** Round to elongated holes caused by the generation or accumulation of entrapped gas or air.

The usual cause of blow holes is sand rammed too hard (decreasing the permeability), permeability too low for the job, sand and core too wet (excessive moisture), insufficient or closed-off core vent, green core not properly dried, incompletely dried core or mold wash, insufficient mold venting, insufficient hydrostatic pressure (cope too short), cope bars too close to mold cavity, wet gagger or solder too close to mold cavity or poor grain distribution. Any combination of hot and cold materials which would lead to condensation as a hot core set in a cold mold or visa versa. A cold chill or hot chill will cause blows, also wet or rusty chaplets.

● **Blister.** A shallow blow covered over with a thin film of metal.

● **Pin Holes.** Surface pitted with pin holes which may also be an indicator of sub-surface blow holes.

● **Shrink Cavity & Shrink Depression.** These defects are caused by the lack of feed metal causing a depression on the surface of the casting, a concave surface. The shrink cavity, a cavity below the surface but connected to the surface with a dendrite crystal structure.

- **Cold Shot.** Where two streams of metal in a mold coming together fail to weld together. This defect is usually caused by a too cold metal poured too slowly or a gating system improperly designed so that the mold cannot be filled fast enough.

- **Misrun.** A portion of the casting fails to run due to cold metal, slow pouring, insufficient hydrostatic pressure or sluggish metal (nonfluid due to badly gassed or oxidized metal).

- **Scab.** Rough thin scabs of metal attached to the casting by a thin vein separated from the casting by a thin layer of sand. Usually found on flat surfaces, caused by hard ramming, low permeability and insufficient hot strength. Sand does not have enough cushion material, wood flour, etc., to allow it to expand when heated. Unable to expand it will buckle causing scabs along with, but not always, rattails, and grooves under the scabs that are also called pull downs. It is the pull downs that bring about the scab.

- **Blacking or Mold Wash Scab.** When the blacking or wash on the mold or core, when heated, breaks away and lifts off of the surface like a leaf and is retained in or on the metal. The cause is a poor binder in the wash, improperly dried wash, poor wash formula or all of the above.

- **Sticker.** A lump or rat (bump) on the surface of the casting caused by a portion of the mold face sticking to the pattern and being removed with the pattern. This problem is caused by poorly cleaned, shellacked, polished pattern, rough pattern, cheap shallac, tacky shellac, sticky liquid parting, cold pattern against hot sand or insufficient draft.

- **Crush.** Caused by the actual crushing of the mold causing indentations in the casting surface. This is caused by flask equipment, such as bottom boards or cores that are too tall or too large for the prints or jackets. Also rough handling.

- **Hot Tears.** Actually a tear or separation fracture due to the physical restriction of the mold or the core upon the shrinking casting. The biggest cause is a too high hot strength of the core or molding sand. These defects can be external or internal. A core that is overly reinforced with rods or an arbor will not collapse.

If you restrict the movement of the casting during its shrinking from solidification to room temperature, it will literally tear itself apart.

- **Gas Porosity.** Widely dispersed bright round holes which appear on fractured and machined surfaces. This defect is caused by gasses being absorbed in the metal during melting. This gas is released during solidification of the casting. Cause is poor melting practice (oxidizing conditions) and poor de-oxidizing practice.

- **Zinc Tracks.** Found on the cope surface of high zinc alloy castings. The defects are caused by the zinc distilling out of the metal during pouring. This zinc oxide floats up to the cope and forms worm track lines on the casting when the metal sets against the cope. The problem is caused by pouring too hot (metal flaring) in the ladle or crucible, pouring the mold too slow or insufficient gates. The mold must be filled quickly before the damage can be done.

- **Drops.** Where a portion of the cope sand drops into the mold cavity before or during pouring. The causes are bumping with weights, rough clamping, weak molding sand (low green strength), rough closing, jackets placed on roughly, etc.

- **Washing and Erosion.** The sand is eroded and washed around in the mold, some of which finds its way to the cope surface of the casting as dirt sand inclusions. It can come from the gating system or in the mold cavity. The causes are a too low hot strength, a too dry molding sand, poor gating design, a deep drop into the mold, washing at the point of impact, metal washing over a sharp edge at the gate, metal hitting against a core or vertical wall during the pouring.

- **Inclusions.** Dirt, slag etc. This defect is caused by failure to maintain a choke when pouring, dirty molding, failure to blow out mold properly prior to closing, sloppy core setting causing edges of the print in the mold to break away and fall into the mold. The drag should be blown out, the cores set and blown out again. Dirt falling down the sprue prior to the mold being poured or knocked in during the weighting and jacketing. For the most part, it's just dirty molding.

Chapter 2
Mold Making

Foundries throughout the world are rapidly turning to new types of sand molding to improve casting quality, productivity, profits and working conditions.

In the casting process a pattern is encased in an aggregate of materials to form a mold; this material is called the *investment*. The word is derived from a term once used to describe special apparel, such as a cloak or uniform. Thus, the pattern "cloaked" in material is called an investment mold.

INVESTMENTS

Two basic investments are used for casting. For nonferrous metals cast at no more than 2000° F, the most commonly used investment is calcium suphate semihydrate: plaster of paris. It is mixed with various *refractories*. Refractories, such as calcined silica or talc, are heat-resistant materials. Some investment forms of plaster of paris contain fibers for added strength in both the "green" and dry states; they can be fiberglass or ceramic fibers. Also, ingredients are often added to promote permeability.

Ferrous and nonferrous metals intended for casting temperatures up to 3100° F are used with hydrolyzed ethyl silicate, a binder, combined with various refractory material, such as zircon, calcined mullite or fused silica, to which is added a small amount of gelling agent (usually a 7.5 percent ammonium carbonate solution). Hydrolyzed alkysilicate is also used as a binder in some formulas.

A phosphate-bonded refractory investment if frequently used a backup mix for flask-molded patterns which have been precoated with a hydrolyzed ethyl silicate investment. Common investment formulas and their applications are given in Table 2-1.

Investment formulas for nonferrous metals are many, but basically they consist of a binder and a refractory, often combined with various items for increased strength and permeability.

- Binders include:
  Plaster of paris
  Portland cement
  Lindseed oil
  Clay
  Ethyl silicate
  Phosphates
  Resins
  Silicate of soda
- Refractories include:
  Wollastonite
  Silica sand
  Brick dust
  Fiberfrax
  Chrysolite
  Zircon sand
  Olivine sand
  Mullite
  Ceramic grog
  Asbestos
  Vermiculite
- For strength, one of the following can be added to the investment:
  Fiberglass
  Ceramic fibers
  Chicken wire
  Hardware cloth
  Metal rods (gaggers)
- Permeability can be increased with:
  Foaming agents
  Wood flour
  Sea coal (crushed coke)
  Perlite
  Course grog
  Luto (reused plastic of paris binder)

Table 2-1. Investments and Their Applications.

Commercially Prepared Investments	Mold Construction	Casting Metals	Max Pouring Temp
Hydro Perm (U.S. Gypsum Co.)	Flask mold Full mold	Brass Bronze Aluminum	2000°F
Investment R Investment R 555 (Ransom & Randolph Co.)	Flask molds Hand-built molds	All (below max pouring temp(	2000°F
Duracast 20 (Derr Mfg. Co.)	Flask molds Hand-built molds	All (below max pouring temp)	2000°F
No. 1 molding plaster (Georgia Pacific Inc.)	All	All (below max pouring temp)	2000°F

Noncommercial Investments	Mold Construction	Casting Metals	Max Pouring Temp
95% talc 4% Hi-Early cement 1% asbestos (by weight)	All	Brass Aluminum Bronze	2000°F
silica sand and fireclay (1:1 by volume)	Full molds	All	2000°F
70% Nol 1 molding plaster 29% talc 1% hydrated lime 0.34 portland cment (by weight)	All	Brass Bronze Aluminum	2000°F
54% .100-mesh silica sand 32% No. 1 molding plaster 13% talc 1% Hi-Early cement (by weight)	All	Brass Bronze Aluminum	200°F
200-mesh olivine flour and No. 1 molding plaster (3:2 by volume)	All	Brass Bronze Aluminum	2000°F
Calcined plaster of paris (gypsum)	Flask molds Full molds	aluminum	(none)
Plaster of paris and silica flour (1:1 by volume)	All	Brass Bronze Aluminum	2000°F
Brick dust and plaster of paris (3:2 by volume)	All	Brass Bronze Aluminum	2000°F
100-mesh silica sand and Hi-Early cement (10:1 by volume)	Flask molds Full molds	All	2000°F

String
Paper
Coke breeze

Some investment molds are made by casting the pattern in an investment of one kind, and then making the remainder of the mold with another. These are called *backup* investments. The initial coating is called a *precoat mix*.

The precoat mix provides maximum detail, refractory properties and texture. The backup mix gives the mold strength and necessary permeability. Because the metal does not come in direct contact with the backup mix, it can be made of coarser and cheaper materials, reducing the overall mold cost. This practice is widespread in the manufacture of nonferrous castings, as well as for full ceramic molds. Commonly, the pattern is coated with a good commercial investment to a thickness of ¼-inch, and the remainder of the mold is made with one part 60-mesh silica sand to one part plaster of paris, by volume. Some fibers are added strength, and wood flour for permeability.

Another common backup mix is used the same way is 60 percent vermiculate, 30 percent plaster of paris and 10 percent asbestos.

In ceramic investment molds using an ethyl silicate bonded refractory precoat, the backup mix often contains a phospate binder to conserve a cost.

How to finish the surface of a cast bronze is a matter of relative considerations. A rough finish on large bronze would be more acceptable that on a smaller piece. The purpose for the piece and its statement as a design will determine its finish also. The final surface treatment has a bearing on the investment method chosen. For example, a bronze that is going to be polished and buffed over its entire surface could be invested in a simple, cheap and foolproof cement-bonded sand mold: 10 parts sand to one part cement.

Whether one should compound an investment or purchase one commercially prepared is a matter of choice and conditions. By and large, a commercial investment will produce more consistent results. The initial cost is often lower than for one you could compound yourself because manufacturers of investments purchase the ingredients in large quantities. They also have the advantage of facilities to manufacture the product under careful quality control. On the other hand, if you are a great distance from a source of raw materials, freight cost becomes a factor. Also, if you are a consumer of small quantities of investments, casting will be more costly to you; small lots of commercial investments are expensive. In some cases, certain materials are unavailable in small lots because it is not profitable for a supplier to break a bulk lot. If your location and financial position is such that you cannot get what you want, but you can buy, say, plaster of paris and some bricks locally, this is the route to take.

Before you decide that the investment you're using is not performing correctly, carefully analyze your complete operation—pouring, calcining, etc. Often, after a series of bad or unsatisfactory castings from a given investment, the investment was blamed and a new one tried, only to have the trouble persist, proving the fault to be elsewhere. I have seen beautiful, flawless castings made from cheap homebrew investments, and some real bad castings that came from complicated, expensive investments. The point is clear: each stage, from the pattern through the final patina, must be carried out with care to achieve good results.

The end use, design, size, surface treatment, composition and pouring temperature of a casting are all factors that have a definite bearing on the choice of investment and method you use.

DIRECT CASTING INVESTMENT MOLD

The direct investment casting mold is one in which the core, pattern and mold are all made progressively, in that order, to produce a *one-time* or *unique* casting. The first step is to build an *armature* or framework, of steel or soft iron wire.

The core, which will resemble the general shape of the casting, is modeled on the armature (Fig. 2-1) with investment mixed thickly in small amounts and progressively built-up with spatula and hands. The surface is kept damp so that each succeeding application of investment adheres to the last.

The next step is to build up the final shape with victory wax in the thickness desired for the casting (Fig. 2-2). The wax is applied in small patties or balls, welded with a warm modeling tool. Make the

Fig. 2-1. The core for a mold can be built up directly on a metal framework called an armature.

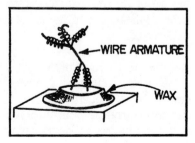

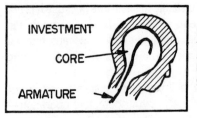
Fig. 2-2. The final shape and wall thickness of a hollow mold built around an armature is determined by the wax layer over the investment core.

wax as uniformly solid as possible. The surface of the wax is then detailed as desired, and nails are inserted through the wax and into the core. These will support the core after the mold has been dewaxed, and the necessary gating work attached.

FOAMED PLASTER INVESTMENT MOLDS

These molds are smooth with air cells just below the surface. During setting and subsequent drying, the air cells become interconnected, permitting gases to escape when the metal is poured.

Just about any conventional formula used for regular investment molds can be used by adding a foaming agent.

Foamed, highly permeable molds can be produced with three parts silica flour and two parts No. 1 molding plaster mixed with any dishwashing liquid.

The pattern and gating is prepared in the conventional manner and filleted down to a shellacked molding board. The pattern is then given a thin coat of investment (unfoamed).

A flask or sleeve is placed around the pattern and carefully secured to the molding board.

A rubber-disc mixer can be driven with a hand drill or drill press. The usual investment mix is one part water to one part dry investment.

The desired weight of water is placed in a clean mixing bucket, and the correct weight of investment is added, followed by a small squirt of dishwashing liquid. Then the mix is allowed to stand for 30 seconds or so. The mixer disc is started and lowered to within 1 or 2 inches of the bottom of the bucket and the investment is mixed until well wetted.

Then, the disc is raised to a point at which a vortex is created that will make a foam of the investment.

The investment, as it foams, will increase in volume by as much as 70 percent in 60 seconds. The trick is to lower and raise the disc to control the size of the air cells. A very fine, even cell structure along with approximately a 50 percent increase in volume signals the end of the process. After the flask has been filled with the foamed mixture, it is jiggled slightly to dislodge any big bubbles that were formed during pouring.

Some commercial investment mixes contain a foaming agent, making the agitation described unnecessary. When making molds from these mixes, it is important that enough is mixed each time to completly fill the flask. A fresh batch will often not bond to the previous dry batch.

Foamed investments can be used to produce full molds with expendable patterns, or two-part flask molds using flexible rubber patterns or rigid, properly drafted patterns.

CORED MOLDS

Any bronze with a section much thicker than ¾ inch to 1 inch should not, if at all possible, be poured solid. Aside from the weight of the piece, problems with shrinkage, risering, chilling and a rough surface caused by heating the inner wall of the mold cavity to excess, will be encountered.

A core is a body of investment, or refractory aggregate, used to make a casting hollow. Its external shape becomes the internal shape of the casting.

The process for a cored bronze is the same used for a solid bronze, with one exception: the pattern for a cored bronze must be hollow. Small galvanized steel nails are pushed halfway through the wall of the hollow wax pattern. The cavity of the wax pattern is filled with the same kind of mix used for a backup investment, and provided with a wax or string vent for gas that will escape from the core when the metal is poured in around it. The mold for the casting is produced like any other bronze mold. When the mold is dewaxed and calcined, the nails (embedded in both the core and the mold wall) support the core and maintain its correct position in relation to the

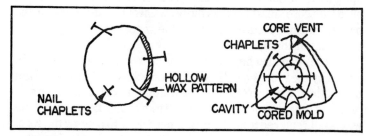

Fig. 2-3. Chaplets (support nails) are used to hold the mold core within the final investment once the wax has melted away.

31

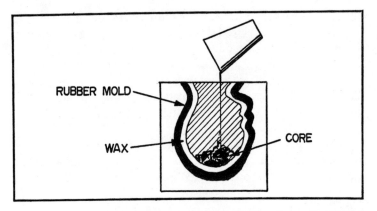

Fig. 2-4. The rubber mold that formed the wax pattern can be used to hold the pattern while the core is poured.

mold wall (Fig. 2-3). Nails used in this manner are called *chaplets*. After casting, the chaplets are cut off and chased down to the level of the outer surface of the casting—or they can be punched out and the resulting holes stopped with brass plugs.

The core investment should have sufficient permeability, and be vented to the outside of the mold. The core is, in most cases, nearly completely surrounded by metal. The gases generated during pouring must be free to get out of the core through a vent. Otherwise, the gases will have to exit through the surrounding liquid metal causing holes to develop.

Because the texture of the internal core surface is of little consequence in most cases, and permeability is desirable, the core investment is made of course materials and contains some wood flour and coral. In some cases, it is made of a foamed, Hi-Perm investment.

Often, large or extremely tall hollow wax patterns are held upright on the surface of water in a container while they are being filled with core investment; as the investment fills the interior of the pattern, it sinks progressively into the water, preventing the pattern from distoring.

Cores can also be poured in the hollow wax while it is supported by the rubber mold in which it was slip-cast (Fig. 2-4). With this method, the chaplets are located after the wax, with its core, is removed from the rubber mold. The disadvantage here is that if the wax is not removed from the rubber mold prior to pouring the core, there will be no way to determine whether the wax has any thin spots or defects that would require fixing from the inside. Once the wax has been removed from a rubber mold for inspection or further work, there is no way to return it properly to the rubber mold for pouring the core.

Hollow castings might also be constructed by first placing the pattern on a molding board, running the chaplets into position, and completing the gating system as far as possible down to the board (Fig. 2-5A). The pattern is then coated with investment, and the *major mold body* of backing investment is fashioned down to the board (Fig. 2-5B).

When the mold has set, it is removed from the board, turned over and supported within a box (a crate would do).

Nails are driven around the "top" of the mold projecting from the box, and the gating system is completed. (Fig. 2-6). A paper sleeve is placed around the mold and extending to the level of the pouring basin. It is sealed with wax where it joins the top face of the mold. The core investment is then poured into the pattern(Fig. 2-7) up to the top edge of the pouring basin, completing the mold. Before it sets completely, a welding rod is used to vent the core in several places.

Sometimes a core is in an area that cannot be filled through a large opening. The horse in Fig 2-8A represents such a problem. The way to accomplish the pouring, in this case, is to cut a ½-inch

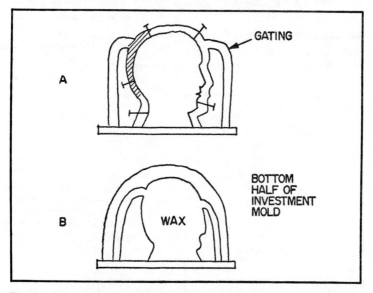

Fig. 2-5. Some molds can be partially made on the molding board. After the pattern is secured and the chaplets installed (A), a half mold is constructed (B).

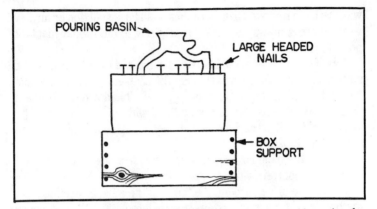

Fig. 2-6. The half mold, built on the molding board, is inverted into a box for support so that its gating can be completed.

hole in the bottom of the belly. Chaplets, to support the core, are placed in position along with a waxed string vent. The horse is turned over on its back, and the cavity is filled completely with investment using a bulb syringe (Fig. 2-8). When set, the mold is completed normally.

A sleeve around the bottom of a pattern can be used to form a substantial base on the end of the core, and in cases when the core configuration is simple and its bulk sufficient, chaplets are unnecessary. The print (Fig. 2-9) will support the core and keep it from moving during pouring. When the core is set, the sleeve is removed, and a groove is cut midway around the print that will act as a lock to prevent vertical core movement after the mold proper has

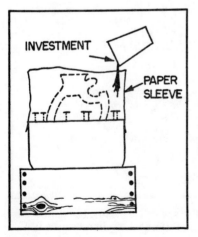

Fig. 2-7. A paper sleeve is placed around the half mold and the investment is poured in to the level of the pouring basin. This process completes the mold.

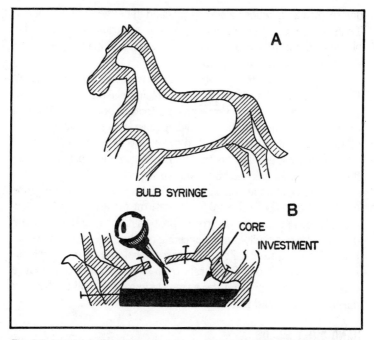

Fig. 2-8. A small, hollow pattern for a horse (A) would not have a large opening through which it can be filled with core investment. A hole must be made, and the investment injected (B).

been made. The pattern is set upright on its print, gated and then the mold is completed in the usual manner.

Avoid cores that require very thin sections; they will be fragile, and cannot be supported properly. Also be careful to prevent stress

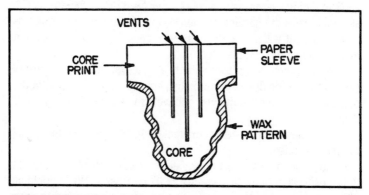

Fig. 2-9. The core print will support simple cores within the final mold, making chaplets unnecessary.

due to shrinkage where a core is pinched in between two cavities. To prevent metal from closing off a vent and causing a hole to be blown through the casting, fit a small ceramic sleeve or steel tube through the vent to hold it open.

Some recommended core mixes are: Hydro-Perm made by U.S. Gypsum Inc; two parts No. 1 molding plaster to three parts 70-mesh olivine sand to three parts luto; and one part 100-mesh sharp silica to one part No. 1 molding plaster to two parts perlite. (The ingredients for the last two mixes are added by volume.)

Most commercial investments containing a few parts of wood flour or coral for permeability will work well.

Half-inch chopped fiberglass strands can be added to any core mix (other than Hydro-Perm) to increase the dry strength of a core.

Never use copper or bronze boat nails for chaplets because they will oxidize badly during calcining and become thin and weak. Use galvanized-steel nails or, for extremely large cores, stainless-steel nails. Steel rods (gaggers) are used to reinforce some large cores, but always present the hazard of expanding and contracting during calcining and fracturing the ore.

HAND-BUILT MOLDS

Assuming that the pattern has been fitted with a gating and pouring system, we are going to construct the investment mold, step by step.

Start to construct the investment mold proper by taking a ¾-inch plywood board big enough to work on, and give the surface two coats of orange shellac. Let it dry to a hard coat.

You are now ready to start the mold. Wash the entire pattern and the gating system with alcohol, then let it dry completely. The alcohol lowers the surface tension so that when wet investment is applied to it, it will not bead up or leave uncoated areas. It will coat evenly and cover completely.

Next, attach the pattern with its gating system to the center of the board, upside down, by running a hot tool around the top of the pouring basin to melt some wax and weld the whole assembly down (Fig. 2-10).

Silica flour and plaster would make a good investment material for this exercise.

Add enough chopped fiberglass to a dry mix of three parts 200-mesh silica flour (a refractory) to two parts No. 1 molding plaster (a binder) so that it is distributed evenly throughout the mix (¼ ounce of ½-inch chopped fiberglass to a 3-pound batch of silica

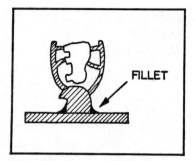

Fig. 2-10. A gated wax pattern, suitable for a hand-built mold, is first secured to the molding board with a fillet of wax fashioned from the pattern's pouring basin with a warm wax tool.

FILLET

flour and plaster should be enough). Mix the dry ingredients well and place the mixture in a dry tin can.

In a shallow container, put a small amount of water (¼ cup or less). To this water, slowly add the dry investment mix, stirring with a ½-inch brush until the mix is creamy enough to adhere to your wax pattern without running excessively.

Working from the bottom up, coat the entire assembly, taking pains to avoid bubbles. Work the investment very carefully into all details and pockets so that no air spaces are left. Recesses may have to be stippled with the brush to be filled.

Because investment will begin to set once it is added to water, work with small amounts at a time, applying it quickly.

When the investment in your mixing container gets too hard to handle, throw it out, clean the container and start a new batch.

Allow each coat to dry firm—but still damp, before applying the next. Paint on coats until you have covered the entire wax pattern and gating assembly to a thickness between 3/16 inch and ¼ inch. Now allow the entire assembly to set—but not completely dry.

The backup investment I recommended at this point is, by volume, one part coarse 50- to 60-mesh silica sand—washed and dried—to one part No. 1 molding plaster. It is all mixed very well together with the chopped fiberglass.

The backup mix is used to make up the bulk of the mold and requires more permeability. Because it is thicker than the painted-on coat, use a coarse silica sand in place of flour.

Apply the backup mix to the pattern with your hands, starting from the board and working up, after having mixed in water as you did earlier, but for a thicker consistency. Avoid overlaps and bubbles.

The thickness of the backup investment depends on the size of the casting you are making; 1-inch sides being the minimum for small to medium-sized patterns, and from 6 to 8 inches all around for larger patterns. Play it safe; don't try to be too conservative—you might lose the casting.

CEMENT MOLDS

Cement investment molds are made in either two-part, cylindrical metal flasks or in rosin-paper flasks.

The investment cannot be poured; it must be tamped down around the pattern. The investment consists of 10 parts (by volume) of 100-mesh silica sand to one part Hi-Early cement.

The pattern should be made of three parts pine rosin (by weight) to one part victory wax. The cement and sand are mixed dry, and then tempered on the heavy side—8 percent moisture.

The pattern is imbedded in 4 or 5 inches of cement and sand, and tamped down in the bottom of the flask. Then the mold is progressively built-up by tamping to the required height and leveled off. The mold is allowed to set overnight, then inverted and de-waxed at 600°F. The cement mold does not require calcining at all, only dewaxing. The mold, when cold, can be filled with any metal or alloy up to 2800°F in its molten state. Large cast-iron fountain and architectural pieces are often made this way.

For large gray iron and bronze pieces, when surface smoothness is a factor, the wax pattern can be embedded within the drag half of a two-part flask. Then it is parted (generally, at midpoint) and the half-mold is allowed to set overnight. The parting plane is coated with 10-weight motor oil, and the cope half is made and allowed to set. The mold is clamped together and dewaxed. After dewaxing, the mold can be separated. The entire cavity, which at this point is exposed, is washed with a core wash, such as Zirc-O-Graph mixed with naptha. The naptha is ignited and burns off, setting the wash. The mold is then reassembled, clamped together and filled.

If the piece is to be cored, it is lined with clay to the desired metal thickness, and provided with the necessary prints for support. The core is made of the same mix used for the mold. The reason for using a rosin-wax pattern is that in dewaxing a cement mold, you do not calcine; heat is only used to melt the wax out. Thus, some wax is absorbed back into the mold face, making it soft during this operation. However, when the mold cools to room temperature, the rosin will reharden the mold face.

Extremely large castings have been successfully cast in bronze and iron this way, some being 20 tons or more in weight. Cement molds are best poured through bottom gating with several whistler vents on top, because wax left in the mold to ignite and burn must have free passage out.

CERAMIC SHELL MOLDS

Usually, ceramic shell molds are made by dipping the gated wax pattern in a watery mixture (a slurry) of ethyl silicate (insoluable) mixed with a suitable refractory, such as fused quartz sand. Quartz sand is graded into various *meshes*, from 200-mesh (flour-fine) to 30-mesh (very coarse).

With a change in the acidity-alkalinity balance of the ethyl silicate solution, silica precipitates as a gel. The gelling is obtained when the ethyl silicate and alcohol mixture is hydrolyzed by adding water and a catalyst, such as hydrochloric acid. A common ethyl silicate formula is made up of 2666 cc of ethyl silicate, 660 cc of isopropal alcohol, 15 cc of HCL (concentrated) and 4000 cc of distilled water: The ingredients are mixed until completely hydrolyzed, then diluted with an equal portion of isopropyl alcohol (by volume). Use only stainless-steel tools and plastic mixing containers for this solution.

To make up the slurry, add 6.65 quarts of ethyl silicate solution to 22.5 pounds of 200-mesh fused silica flour. Dip the clean pattern in the ceramic slurry and drain off the excess; then stucco it lightly by sifting on a mixture of 50- and 100-mesh silica sand through a riddle. After it dries, repeat the procedure until the proper shell thickness is built up.

Some operators, after two dips and stuccoing, decrease the fineness of the stucco progressively.

Phosphate bonded investments are cheaper than ethyl silicate and are used as a backup for full ceramic molds. The thickness depends upon the size of the casting. The average thickness is ¼ inch, but shells up to ½ inch thick are common. The full-ceramic investment mold is produced by basicallly two methods. One involves making a shell on the wax pattern, placing the encased pattern in a flask, and surrounding it with a phosphate-bonded investment. Or, two-part flask molding can be done in the same manner.

There are several precoat formulas and backup mixes on the market for specific uses. The molten pouring temperature, metal composition and the molding method are the determining factors in their selection.

GREEN SAND MOLDING

In green sand molding, a mold is produced with an aggregate consisting of refractory material and a binder. The two basic ingredients are silica sand, the aggregate and a refractory clay as the binder. Two types of green sand are commonly used: natural bonded molding sand and so-called synthetic sand.

Natural, bonded sand is a mixture of sand and clay, ingredients found in nature, mined and sold to foundries by foundry suppliers. They are graded according to their best application. It is common to find several grades in one mine or digging. A much used natural bonded molding and is called Albany sand—from Albany, New York. It is strip mined in seven grades from No. 00 for very small iron, brass and aluminum castings to a No. 3 for heavy castings.

French sand, as its name implies, comes from France. It is used for fine bronze and aluminum castings; in particular, bronze tablets and plaques.

French sand consists of 16.6 percent clay, and is available in particle sizes from 135 to 170 mesh. Natural bonded molding sands can be found in numerous locations throughout the U.S. and are usually named for the location. Greely sand, good for brass and aluminum castings, comes from a pit near Greely, Colorado. Synthetic sands consist of sharp clay-free silica sand with southern or western bentonite added as a binder.

By selecting the proper base sand, and adding the required amount of binder, any type of molding sand you need can be formulated. For example, 100 pounds of washed and dried silica sand with an average fineness of 140- to 160-mesh, 5 pounds of southern bentonite, 4 pounds of sea coal (ground coke), mixed with 4.5 percent moisture, would yield a good molding compound for light cast iron or heavy bronze. For heavy work, try a mixture of 100 pounds of clean, dry silica sand with an average fineness of 58-to 69-mesh, combined with 1 pound of southern bentonite, 1 pound of western bentonite, 1 pound of pitch, 2 pounds of sea coal, and 5 percent moisture. The basic disadvantage of synthetic sands is that they require a *muller* to mix them and to develop the necessary green strength. It's an expensive piece of equipment.

Natural bonded sands can be mixed with a riddle and cut by hand with a molder's shovel. They carry more moisture for a given bond strength—a disadvantage. However, most bronze and aluminum is cast in natural bonded sands except in highly mechanized operations. The term *green* sand refers to a mold made and poured with sand simply tempered with water.

SKIN DRIED SAND MOLDS

The skin-dried mold is produced like a green sand mold with one exception: prior to closing the mold for pouring, spray every interior surface of it with a mixture of 10 parts water to one part molasses or lignin sulphite. Dry the sprayed areas with a torch using a low, broad flame kept in motion to dry the sand slowly, and to avoid excessive heat that might scorch the surface. You want to simply drive out the surface moisture and supply enough heat to set the molasses or lingnin binder.

Dry the mold surface to about ⅛ to ¼ inch in depth, leaving a smooth, hard skin.

Although this system seems quite simple, it takes quite a bit of skill and practice to do properly. This kind of mold is commonly used for tablets, grave markers and highly ornamented castings. This method prevents small sand protrusions on the mold surface from being washed away when the mold is poured. A skin-dried French sand mold will produce a casting with a surface finish superior to that of an investment casting.

Skin-dried molds are best made in metal flasks—wood flasks might burn. After the mold is carefully dried, allow it to cool and then close it. Pour the mold shortly after closing it; otherwise, the moisture in the sand behind the dried surface will eventually migrate to the surface, defeating the purpose of the whole operation.

DRY SAND MOLDS

Dry sand molds are made in steel flasks. The green sand is mixed with a binder such as pitch, linseed oil, lignin sulphite or a commercially made one. Make the mold in the usual manner (as in green sand casting) and, when finished, spray them with molasses water and dry them overnight at 350° to 400°F. Dry sand molds are usually made when thick metal sections are anticipated, and a tough mold is required. Steel castings are commonly poured in dry sand molds. A typical dry sand formula for heavy work is 100 pounds of sand, 3 pounds of plastic fireclay, 3 pounds of bentonite and 2 pounds of pitch.

NO-BAKE MOLDS

No-bake molding involves the controlled mixing of two or more chemicals with sand to form a filled mold which cures or air hardness within minutes at room temperature. No fuel for baking or curing is needed. One of the chemicals, a catalyst or hardener, causes the resin binder which coats the sand grains to undergo a chemical reaction. As time passes, the sand grains are bound together and the mixture sets up or *cures*.

Variables such as temperature, humidity and the reaction rate between the liquid binder and catalyst affect *work time* and strip time. Work time is the period during which the sand mixture can be used to produce satisfactory molds. Strip time is the time after mixing when the pattern can be removed. In any case, no-bake molding offers the foundry numerous advantages. These advantages can be most fully realized when the no-bake process is combined with an environmentally designed, energy-efficient system that can process the sand for re-use. In this sense, *sand reclamation* is the logical extension of no-bake molding.

Proven Advantages

Energy and Time. Significant saving in the cost of thermal energy is achieved because baking or curing is eliminated or greatly reduced. As fuel shortages become more acute and costs rise, this will become even more important. Short, controllable cure times and simpler, more efficient molding procedures also permit high-production schedules to be met. The no-bake process is ideal for producing a wide producing a wide variety of casting sizes where metal pouring flexibility is required.

Sand and Equipment Costs. Less sand is needed because the sand-to-metal ratio for no-bake molding is lower than that for green sand. The advantages of flaskless molding are important. The cost of flasks, handling and maintenance is often eliminated. Casting removal and cleaning are also easier. In addition, the overall capital investment for equipment is usually lower.

Casting Quality. More accurate, truer-to-pattern reproduction improves casting quality. With no-bake's rigid wall molds, less scrap loss and higher yields are often recorded. These dimensionally consistent castings require less cleaning, finishing and machining set-up time.

Economics. Just as in green sand molding, approximately 80 percent to 90 percent of the reclaimed sand from a no-bake mold typically is reusable. Casting producers can stretch supplies of dwindling and ever more costly virgin molding sands. At the same time they significantly-acceptable disposal.

Total System Approach

To gain maximum benefit from no-bake molding, a *total system* includes:
- Automated, continuous sand mixing
- Mechanized mold and core handling

- Controlled metal pouring and cooling
- Casting removal, cleaning and sand reclamation, ideally accomplished by an environmentally designed, single-step system
- Closed-system sand return

Before embarking on no-bake molding in a large operation, foundrymen must consider several factors. These factors include the type of metal; weight and size of casting; production requirements per pattern; type and cost of sand and binder; sand-to-metal ratio; type of mixing equipment; casting cleaning; and the sand reclamation system.

It is common practice with large flask molds to face the pattern with a no-bake mix, then back it up with system sand.

In no-bake molding the internal cores are usually made with the same sand mix. In essence a no-bake mold is a core mold.

NO-BAKE BINDERS

Both organic and inorganic binder systems are available for no-bake molding. In order to coat individual sand grains readily, binders are applied as individual liquid resins. Depending upon the binder system, the hardening additive or catalyst can be introduced into the sand mixture in the form of a liquid, a finely divided solid or a gas.

Organic Binders

Furans. The furan series is the oldest of the no-bake systems and sometimes is called *acid no-bake* because an acid acts as a catalyst. The series exhibits high strength and exceptionally well-elevated temperature properties. Cores and molds can be poured four hours after they have been stripped. Furan binders are available with different levels of nitrogen and water. The grade of binder varies with specific use. Steel producers usually want very low or zero nitrogen and a water content less than 1.0 percent. On the other hand, most gray iron producers prefer 4-5 percent nitrogen and 12-20 percent water. The cost of the binder generally is related to the nitrogen and water content. The lower the nitrogen and water, the higher the cost.

Oil-Urethane (Alkyd). This binder can be used as either a two-or-three-part system. The binder portion is an alkyd resin. The catalyst consists of two parts: a polymeric isocyanate and a metallic dryer (a lead or cobalt compound in liquid form). The purpose of the metallic dryer is to speed up the reaction.

The oil-urethane system is among the least sensitive to sand temperature and moisture variables. Mold stripping and release properties are excellent. The system contains no water, is low in nitrogen and does not produce unpleasant fumes.

Before cores and molds can accept molten metal, they need to air cure for 8 to 12 hours after stripping. To allow quicker pour-off, some users apply heat in a post-bake cycle.

Phenolic-Urethane. This three-part system consists of a phenolic resin, a polymeric isocyanate and a catalyst to regulate curing speed.

Inorganic Binders

Silicate Types. Two methods are used to harden silicate no-bake binders. In the first method, sand coated with sodium silicate is exposed to carbon dioxide (CO_2) gas. The reaction produces a silica gel which binds the sand particles together. The other method employs an ester setting additive to air harden the coated sand mixture.

Silicate-type binders have high hot strength and good ecological properties with low gas and low or no door, stains or toxicity. However, the curing process is sensitive to sand impurities, moisture and temperature. Low internal strength, poor handling properties and poor shakeout are also characteristic.

Phosphate Type. This zero-nitrogen, zero-silicate no-bake system is composed of a phosphate solution and an inorganic powdered hardener to control hardening speed. Strip time can be varied from 20 minutes to 1 hour. Pouring can begin 1 hour after strip.

Sand mixes are clean. Handling and gas problems are minimal. Tucking and compacting may be necessary in corners and pockets since flowability is less than that found in organic sand mixes. Also, sand impurities may alter the work time-strip-time relationship. Shakeout properties compare with those of organic no-bake binders. The nitrogen content is low and there is no water.

Speed, control of work time, strip time and economically favorable low binder-level requirements make the phenolic-urethane system especially attractive for high-production operations. Intervals of as little as two minutes have been reported from filling to strip. Castings can be poured within one hour after stripping. Sand impurities, however, can affect cure rate and strength. During casting some fuming may also occur.

A typical furan mix would be binder 2 percent of the weight of the sand and the catalyst 25 percent of the weight of the binder. The mix could be 100 pounds sand, 2 pounds binder and ½ pound catalyst. The catalyst most often used is phosphoric acid.

Chapter 3
Molding Sands

When we speak of sand casting the first thing that comes to the minds of most people is the sand on the beach or the desert. There are many other places to find sand.

You can prospect for your own molding sands by looking for deposits along creek beds, river banks and cliff as well as the beach and desert. The typical molding sand for foundry use resembles ordinary sand, usually screened to a definite size fraction, mixed with 10 to 20 percent clay to act as a binder to hold the sand grains together. Molding sand is ordinarily quite inexpensive. In fact, natural molding sands are found in many areas which contain the correct proportions of sand and clay, and it is only necessary to dig the material out of the ground. Very few natural sands possess the desired properties in all proportions, however, so it is usual to add other ingredients to improve them.

MOLDING SAND PROPERTIES

Molding sand must possess several characteristics. First, it must be cohesive so that the individual grains stick together while the pattern is being removed, otherwise the mold would break apart. Second, it must be porous enough so that gases and water vapor can escape when molten metal is poured into the mold. To a certain degree, the properties of cohesion and porosity work at cross purposes. The addition of clay or clay-like material will improve the cohesiveness of the sand grains, but at the same time will tend to reduce porosity. Molding sand must also be refractory to withstand the high temperature of the molten metal.

The size and shape of the sand grains influence the properties of the molding sands. Rounded grains such as beach sands do not cohere nearly as well as sharp, irregular grains which interlock and provide a stronger structure when rammed into the mold. Sharp grained sands, therefore, are to be preferred as less clay is required for bonding and the sands will be more porous.

Your local foundry supplier should be able to furnish claybonded moldings sands of the type described in 100-pound bags. Molding sands of this type are usually shipped dry, and before use it is necessary to "condition" them by adding water until the sand develops the correct adhesion. The conditioning process is best carried out by sprinkling the sand with water as it is turned over with a trowel or shovel. Enough water is added so that a handful of the sand compressed by clenching the fingers will stick together like a snowball, leaving a distinct pattern of the fingers and lines in the palm. Excessive water should be avoided. Molding sands of this kind can be used over and over, conditioning whenever necessary, although they will eventually "wear out" and should then be discarded.

To cast sand we make a mold around a pattern, open the mold, remove the pattern, close the mold and fill the cavity left in the sand with molten metal. When the metal has solidified we shake out the mold and have an exact duplicate in metal of our pattern. Anyone who has built sand castles on the beach can attest to how fragile they are. But if we took our beach sand and mixed enough clay with it to give each grain a coating of clay, which when damp is sticky, we would soon realize that we could make great sand castles, not nearly as fragile as our beach castles, which depended upon water alone as a medium to bond the grains together.

With additional experimentation we would find that a mold made of beach sand, clay and water could be used to hold molten metal. The next question that would come to mind is why when the mold is filled with hot metal it doesn't crumble or explode because of the moisture content. Very simple, this is what happens; as the molten metal enters the mold cavity the radiant heat from the metal dries the mold material in advance of the metal flow. The moisture is changed to steam and moves out of the mold through the mold walls because of the porous nature of the molding sand.

Permeability

The ability of the mold material to allow the steam to pass through the walls is called permeability. Permeability can be measured with a meter which measures the volume of air that will pass through a test specimen per minute under a standard pressure.

Some instruments are designed to measure a pressure differential which is indicated on a water tube gauge expressed in permeability units.

In this book we are for the most part concerned with natural bonded sands, to be used in green sand molding. A natural bonded sand contains enough bonding material that it can be used for molding purposes just as it is found in the ground.

Natural molding sands contain from 8 to 20 percent natural clay, the remaining material consists of a refractory aggregate, usually silica grains.

Any natural sand containing less than 5 percent natural clay is called a tank sand and is used for cores or as a base for synthetic molding sand.

Commercial molding sands mined by various companies usually acquire the name of the area where they are mined. The most popular natural bonded molding sand is called Albany, and is mined in several different grades by the Albany Sand & Supply Co., Albany, N.Y. The origin of this sand is from the pleistocene ice sheet of approximately 20,000 years ago which swept down from the north and completely overran what is now known as the Albany District. The result after eons is a seam of fine molding sand approximately 15 inches thick directly under an overburden of 8 inches of top soil.

Before discussing the prospecting of molding sand, let's look at a few sand characteristics necessary for the production of castings of various metals and sizes.

Light gray iron
    Fineness..175
    Clay 12%

Moisture..7.4%
    Permeability...15
    Green compression...4.0

Medium gray iron
    Fineness..111
    Clay 15%
    Moisture..7.5%
    Permeability...40
    Green compression...4.0

Heavy gray iron

  Fineness 73

  Clay 18%

  Moisture...7.6%

  Permeability...70

  Green compression...5.0

Heavy brass

  Fineness..108

  Clay 12%

  Moisture 7%

  Permeability...51

  Green compression...4.0

Light to medium brass

  Fineness..218

  Clay 13%

  Moisture 8%

  Permeability...18

  Green compression...4.0

Aluminum

  Fineness..232

  Clay 19%

  Moisture 8%

  Green compression...5.0

Notice the two differences between the sands; the grain fineness and the permeability required in a sand used to make a mold for a gray iron casting compared to the fineness and permeability required in a sand for use in making a mold for aluminum.

Fineness

  This is a measure of the actual grain sizes of a sand mixture. It is made by passing a standard sample, usually 100 grams, through a series of graded sieves. About 10 different sieve sizes are used. As most sands are composed of a mixture of various size grains there is a distribution of sands remaining on the measuring sieves.

  The fineness number assigned to a sample is approximately the sieve (screen) which would just pass the sample if its grains were all the same size.

  When you fully understand the relationship between the required permeability for a given metal and it's pouring temperature and the relationship of grain fineness to permeability, you will be able to establish what sand you need for a given metal and casting size. As an example gray iron is poured at a temperature of 2700 degrees F and is three times the weight of aluminum which is poured at 1400 degrees F.

Although we are primarily interested in natural bonded sands, their use and properties, we will, however, cover some aspects of synthetic molding sands.

Both sands have their good and bad features and the choice depends upon the class of work and the equipment available.

The basic components of most molding sands are silica and a clay bond. However molding sands can also be made up of other types of refractory materials such as zircon, olivine, carbon, magnetsite, sillimanite, ceramic dolomite and others.

Molding sand is defined as a mixture of sand or gravel with a suitable clay bond. Natural sands are sands found in nature which can be used for producing molds as they are found. Synthetic molding sands are weak or clay free sands to which suitable clay or clays are added to give them the properties needed.

You should be aware, however, that in the past 20 years there has been a real crop of sand medicine men who sell all types of additives and dopes for molding sands to cure the foundryman's problems.

Most of these products are simple additives which have been disguised in some manner. Make it a point never to buy any sand additive or product if the manufacturer refuses to divulge exactly what the product is.

We are living in an age of rediscovery and many of these rediscoveries are disguised to look like new products. In order to know and control your sand you must know exactly what you are adding and using. There must be no unknowns, otherwise you are in the dark.

Do not buy any foundry product if the manufacturer will not divulge the contents. This information is vital to the control and understanding of your operations.

Avoid complicated sand mixtures, they only lead to confusion and are most difficult to control. A simple sand with the proper grain size and distribution with sufficient bond and moisture, will give much better results than one which is complex. Complex sands are usually a product of experimentation with various additives trying to accomplish some particular illusive result. This is usually due to insufficient understanding of molding sands, their formulation, limitations and uses.

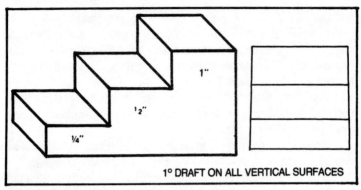

Fig. 3-1. Step pattern.

Keep the types, kinds, number and amounts of sand additives to a minimum for best results.

Avoid the use of products which are sold as cure alls. No product can offset poor practice.

Refractoriness

This is the ability of sand to withstand high temperatures without fusing or breaking down.

From this we can deduce that a sand used for casting steel must be more refractory than one for brass or aluminum because of the greater pouring temperture involved. Also, a sand used to cast large heavy castings must be more refractory than one used for light thin castings of the same metal.

As we are primarily dealing with natural bonded sands the refractoriness of the sands can vary over a wide range. When a naturally bonded sand contains appreciable amounts of fluxing agents (various mineral salts, organic material and oxides) that lower the fusion point of the sand, it may melt or fuse to the casting. There are various costly instruments used to determine the refractoriness of molding sands. In the absence of such testing equipment we must pour samples of various thickness into the sand in question and examine the surface of these test castings for sand fused to the casting. This actual experience will be more useful to you than all the instruments in the world. This is what we call getting your hands in the sand. In order to start our testing we need a wood pattern (Fig. 3-1). This is a step pattern 6 inches long with three different step thicknesses.

We make a mold of our step pattern in the conventional manner with the sand in question and cast it in brass or bronze at 2200 degrees F. Allow the casting to cool to room temperature in the mold, then shake it out (Fig. 3-2).

The cold casting is carefully examined for adhered sand which will not peel off easily with a wire hand brush. Any area on which the sand will not peel off is examined carefully with a magnifying glass. This examination will readily show if the sand has indeed melted or vitrified under the heat of casting and has in fact welded or fluxed itself to the casting.

Let's say that the ¼ inch thick section and the ½ inch thick section peeled nicely and is free of any adhered sand and presents a nice smooth metal surface, but the 1 inch section does not peel clean and shows vitrified sand on its surface. This would indicate that the sand has a suitable refractoriness for light brass and bronze up to ½ inch in section thickness but is unsuitable for 1 inch and thicker sections.

Should all surfaces peel clean then you must work up a new and thicker pattern. By experimenting you can readily determine the limits of the sand in question.

The reason brass or bronze is used for the test and not aluminum is because aluminum pours at such a low temperature. It would be hard to find any sand that would fuse at these lower temperatures. When performing the refractoriness test the mold must be carefully made and rammed properly. A soft mold (under rammed) or a mold which has not been rammed evenly could give

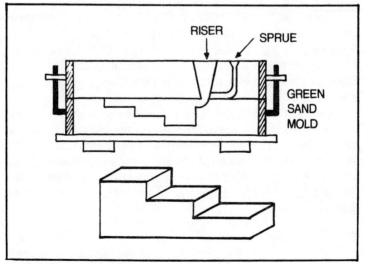

Fig. 3-2. Casting step pattern.

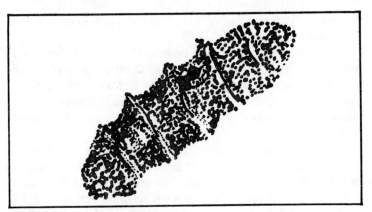
Fig. 3-3. Squeezed sample of sand.

you a defect called penetration which you might falsely identify as poor refractoriness. In this case the surface is rough and contains metal mixed in the sand. But enough of this, we will get into it deeper as we go on.

Green Bond Strength

Green bond strength is the strength of a tempered sand expressed by its ability to hold a mold in shape. Sand molds are subjected to compressive, tensile, shearing, and transverse stresses. Which of these stresses is more important to the sands molding properties is a point of controversy.

Here is one green compressive strength test. A rammed specimen of tempered molding sand is produced that is 2 inches in diameter and 2 inches in height. The rammed sample is then subjected to a load which is gradually increased until the sample breaks. The point where the sample breaks is taken as the green compression strength.

The devices made to crush the specimen are of several types and quite costly. The readings are in pounds per square inch. It sure wouldn't take much thinking or doing to come up with a home made rammer or a compression device. As you are only interested in whether the sand is weak, strong, or very strong, the figures you give for values are only relative and you can call them whatever you like.

Tensile Strength

Tensile strength is the force that holds the sand up in the cope. And, as molding sands are many times stronger in compression than tensile strength, we must take the tensile strength into account. Mold failure is more apt to occur under tensile forces.

Where compression strength is measured in pounds per square inch, the tensile strength of molding sands is measured in ounces per square inch.

The tensile strength, which is the force required to pull the sample apart, is determined very easily.

Pick up a handful of riddled tempered sand and squeeze it tightly with your palm up. Open your hand and observe if the sand took a good sharp impression (Fig. 3-3).

Now grab the squeezed sample with both hands between the thumb and first finger of both hands and pull it apart.

Examine the break. It should be sharp and clean, not crumbly. By observing the break and noting the force required to pull the sample apart you can tell a lot about the tensile strength and general condition of the sand. Make another sample. This time grab it about midway with the right hand thumb and first finger, shaking it until the end that protrudes breaks off. Note how much force was required to make it break (Fig. 3-4).

Dry Strength

A mold must not only hold it's shape in the green state, it must also hold it's shape in the dry state. This is an important property and is measured as dry compression. It allows the test specimen to dry out before testing which is then carried out in the same manner as for green strength. A good average figure is 30 pounds per square inch. Dry strength should be no higher than necessary. Excessive dry strength results in a critical sand. If the molding sand has a too high dry strength it will not give or break down as the casting shrinks during solidification. This will cause hot tearing of the casting.

Durability

Durability is the measure of the ability of the sand to withstand repeated usage without losing its properties and to recover its bond

Fig. 3-4. Sharp clean breaks.

strength after repeated use. The sands fineness and the type and amount of clay bond determines the sands durability. The ability of the bonding clay to retain it's moisture is also an important factor.

Moldability

This characteristic is also related to the nature of the bonding clay and the fineness of the sand.

Because the base sand determines the resulting finish of the casting, it should be selected with care. Keep in mind the type, weight and class of casting desired. The three or four types of screened sands formerly used for a base has given way to the practice of blending one coarse sand with a fine sand which results in a better grain distribution. This has been found to produce a better finish and texture. Each of the two sands selected should have a good grain distribution within itself. Contrary to popular belief, additives of an organic or carbonaceous nature do not improve the finish but only furnish combustibles resulting in better peel.

Blended Sands

Use the following suggested sand blends as a guide when selecting your own basic blends. Start with a good high grade silica, washed, dried, screened and graded. Adjustments can be made to the base after sufficient tracking is done.

- For heavy iron use green or dry sand. Fineness 61 and 50. Permeability 80 to 120.
- For medium iron, green sand.
  Fineness 70 and 45. Permeability 50 to 70.
- For light squeezer iron, green sand.
  Fineness 110 and 80. Permeability 20 to 30.
- For stove plate iron, green sand.
  Fineness 200 and 160. Permeability 9 to 17.
- For heavy green steel, green sand.
  Fineness 60 and 35. Permeability 140 to 290.
- For heavy steel, dry sand.
  Fineness 55 and 40. Permeability 90 to 250.
- For light squeezer malleable iron.
  Fineness 130 and 95. Permeability 20 to 40.
- For heavy malleable iron.
  Fineness 80 and 70. Permeability 40 to 70.
- For copper and monel.
  Fineness 150 and 120. Permeability 30 to 60.
- For aluminum.

Fineness 250 and 150. Permeability 6 to 15.
- For general brass.
  Fineness 150 and 120. Permeability 12 to 20.

PETRO BOND MOLDING SANDS

Although the traditional molding sands will suffice for most routine casting work, far superior results will be obtained for small parts if a synthetic type of molding sand called prepared Petro Bond is used. Petro Bond is a registered trade mark of the National Lead Company. A Petro Bond sand is a mixture of very fine sand, oil, Petro Bond bonding agent and a catalyst. No water is used and none should ever be added. The combination of oil instead of water, which results in less gas generation, and the extremely fine sand, allows surface finish and detail in castings which cannot be obtained with the use of conventional water bonded sands. Surface finish, particularly with automotive trim parts, is very important because many of these parts have to be prepared for plating. Smooth, finely detailed castings save a lot of grinding and polishing operations which are often more expensive and time consuming then the actual production of the castings.

For the amateur home foundryman, what is often a discouraging activity when conventional molding sands are used becomes a real pleasure as the superior results are realized with Petro Bond prepared sand. Prepared Petro Bond sand can be used over and over again except for the tin layer of sand in direct contact with the casting. This layer will be burned black and should be separated from the rest of the sand and discarded. The blackened sand can be reconstituted, but the effort is hardly worth the trouble for small scale operations such as one in your home shop.

Prepared Petro Bond sand is available from selected foundry suppliers in principal cities. It costs several times as much as conventional molding sands, but it is worth the extra cost.

PARTING DUST AND GRAPHITE POWDER

To facilitate the separation of the two halves of the mold and to prevent the sand from sticking to the patterns, a material called *parting dust* is required. Your foundry supplier will stock this material. There is also a liquid parting compound, but it is not recommended for use with Petro Bond sands. About 10 to 25 pounds of parting dust will last the average home craftsman for years so don't buy too much if it can be avoided. Unfortunately, the minimum package amounts of some foundry products are rather more than the typical amateur needs. A sympathetic supplier will "break" a package to sell smaller than minimum shipping quantities. At the same time, buy some graphite powder which is useful for dusting onto patterns to keep sand from sticking to them. Buy only a small quantity as a pound or two will go a long way.

This is a good time to mention that stockmen or warehousemen around foundry suppliers have often been foundry workers at one time. If any one of them takes an interest in what you are doing, he may very well offer some good advice about what material to buy. His advice can be very helpful.

At one time it was common to use fine sand as parting dust, but this practice cannot be recommended due to the health hazard from silicosis. Buy only a "non-silica" type of parting dust.

SYNTHETIC SAND

There is no mystery to synthetic sands whatsoever. The word synthetic is erroneous and misleading. Synthetic sand is not synthetic in any way. It is basically a combination of natural materials provided by Mother Nature, mixed in proportions desired to give certain properties wanted by the user. Natural bonded sand is mixed by Mother Nature and comes as is.

Let's look this synthetic sand business right in the teeth for what it is. Don't start out under the illusion that it is the answer to all your problems and will produce miracles of some sort. It is in reality a very simple and effective sand mixture which when properly prepared and used will give good results. In general it requires no more equipment to test than you should have and use with your naturally bonded sand. It does, however, require a muller or paddle mill to give the correct results. The muller is preferred and is the heart of the system. It is necessary in bonding new and used synthetic sand which requires a kneading and plowing action to develop its properties fully in the shortest time and to produce uniform sand, batch after batch. Also, the muller develops green strength well.

Remember, sand control is the most effective tool you can use to reduce losses and to produce top quality work. Without it you cannot produce consistent results. The lack of sand control by routine testing explains why sand is allowed to become unsatisfactory and difficult to revive. In extreme cases it must be replaced altogether. Even 10 percent of control added to your sand preparation will produce amazing results.

In selecting the base sand, the best rule-of-thumb system is to use the weakest, driest, finest sand you can obtain which will give you the best castings. This can be found only by actual practice.

POPULAR SAND MIXTURES

The following are four of the most popular synthetic sands in use today with nonferrous work:

- Aluminum and brass—Penn wash float silica, AFS fineness 160; southern bentonite, 4 percent by weight; hardwood flour, 200 mesh, 1 to 1.5 percent.
- Semi-synthetic, brass and aluminum—60 percent washed and dried silica, AFS fineness 120, 40 percent naturally bonded sand, Albany O; sufficient southern bentonite to give the desired green strength (7 to 8 psi); hardwood flour, 200 mesh, 1 percent.
- Brass—washed and dried silica, AFS fineness 120; southern bentonite, 4 percent by weight; hardwood flour, 1 to 1.5 percent.
- Copper—washed and dried silica, AFS fineness 130; southern bentonite, 4 to 4.5 percent by weight; hardwood flour, 1 to 1.5 percent; silica flour sufficient to drop permeability to between 40 and 50 if needed.

Held to within reasonable limits of the following values, the first, third and fourth sand mixtures will generally give a higher percentage of good results.

- Mix No. 1: Permeability, 8 to 16; moisture, 3 to 5 percent, as low as possible; green compression, 7 to 8 psi; AFS fineness, 200-160.
- Mix No. 3: permeability, 12 to 20; moisture, 3 to 5 percent, as low as possible; green compression, 7 to 8 psi; AFS fineness, 160 to 120.
- Mix No. 4: permeability, 35 to 50; moisture, 3 to 5 percent, as low as possible; green compression, 7 to 8 psi.
- Mix No. 2 carries about 6 to 7 percent moisture.

For heavy brass and aluminum, vary your sand accordingly for high permeability. Use a coarser sand and hold your moisture as low as possible. Bentonite has its greatest bonding strength with optimum moisture.

It is difficult or even impossible to set down any hard and fast rules. The acid test is in the end results. You are shooting for a sand which will give you sufficient strength, permeability and lowest moisture, yet produce a food finish.

Permeability can be controlled with silica flour which closes up sand, and with coarse sand which opens it up if the moisture remains constant. Bentonite will control the green strength if the moisture remains constant. Wood flour gives a cushion to prevent over-ramming. It improves the finish, shakeout and collapsibility as well as provides a reducing atmosphere in the mold cavity. The pickup of core sand the butts will tend to open the sand up. Fines will tend to close it up. The trend will show readily from daily sand tests and casting inspection. For some unknown reason, mix No. 1 is preferred in the eastern and southern sections and mix No. 3 is coming into wide use along the west coast. Because of their low moisture and high permeability, they can be rammed very hard and will produce castings with a nice peel that are close to size. Either of these sands can be rammed like bricks. They can be sprayed with a suitable binder such as a lignin sulphite binder and water. The skin can also be dried with excellent results on special work. No mention of sintering point is made since most base sands are above 2300°F and are suitable for nonferrous work. Core sand or sands should be as close as possible to the same fineness as your selected base sand so as to have a minimum effect on physical properties as it enters the system. Whether it be washed and dried silica sand, crude silica, lightly bonded sand, beach or lake sand, bank or dune, the base sand should have as wide a grain distribution as possible. Such a distribution will give you your best overall working conditions.

You will find that the permeability and other figures given in the recommended limits for the four mixes are to be used as a guide. Permeability will run higher than the value given due to the low moisture used. The highest permeability possible at which you can get a satisfactory finish is the best point. The wood flour can be deleted but it is very beneficial and most useful tool in control.

Some metalcasters put in a small percentage of cereal bond with or without the wood flour. This is a matter you must decide for yourself. The sand which produces the best results day in and day out is the best mix for you, regardless less of what's in it.

Daily checks and test kept in an accurate and systematic manner easily can be correlated with your practice. The correct amounts of binders, wood flour (or what have you) to be added or left out per batch depends entirely upon your operation—thus the reason for control by testing. You can graph the results of your daily tests, checking continuously as to where your sand is headed to prevent it from falling off too sharply. When it starts off in the wrong direction, you can pull it back before any harm is done.

KEEPING MOLDING SAND IN CONDITION

When starting a job with new sand that is either natural or synthetic, you will find that the castings will improve with use as the system builds up some burned sand and clay and becomes evenly tempered. The sand will peak out and start down hill as core sand and burnt clay build up in the system. In general, with the addition of new sand to replace the sand lost to drag out on castings, your feet, etc., will keep the system in condition unless you contaminate or throw the entire system off kilter by too heavy additions of sea coal, wood flour, pitch, etc. Then you start over. A lot depends upon the ability of your selected sand to take repeated heating and conditioning as well as its ability to recover time and time again (durability).

If your selection of sand is not suited for the weight range and pouring temperature, you are dead at the start. A fine naturally bonded sand for casting light aluminum castings could last for years with good results, but if used it as a mold material for medium to heavy cast iron, it wouldn't last long at all.

Select the correct sand for the work at hand and avoid too many additives. The percentage of new sand added to maintain the heap also depends upon the daily use. You need approximately 10 pounds of sand per each pound of metal poured per day roughly. Keep your sand riddled free from shot metal and trash. Also, keep it tempered properly.

DRY SAND MOLDS

The number one advantage of a dry sand mold is insurance. There is an even greater strength on large molds requiring a large volume of metal. A dry sand mold is free from moisture and you are not troubled with the defects caused by steam and rapid chilling. Make sure that the added cost of going to a dry sand mold is justified by the added assurance of a sound casting.

A skin-dried mold will often suffice in place of a fully dried mold if it is poured soon after the skin has been dried.

Chapter 4
Molding Tools and Equipment

The basic equipment and materials needed to perform casting operations will be discussed in this chapter.

These items are just about the same as any foundry would use, except the amateur's equipment will probably be on a much smaller scale.

TROWELS

Next to the molder's shovel, the most used hand tool is the molder's finishing trowel.

The #1 standard finishing trowel comes in three widths: 1¼ inch, 1½ inches and 1¾ inches. The blade length is standard for any width—6 inches long. The 1½ inch width is standard. The blade tapers from 1½ inches wide at the handle end to 1 inch wide at the end of the blade which is rounded. The handle tang comes up straight from the blade about 2 inches, then is bent parallel to the blade to receive the round wooden handle. The 2-inch rise gives you knuckle room when troweling on a large flat surface. The only difference between a #1 trowel and a #2 is that the #2 finishing trowel has a more pointed nose (Fig. 4-1).

Finishing Trowel

The finishing trowel is used for general trowel of the molding sand to sleek down a surface, to repair a surface or cut away the sand around the cope of a snap flask.

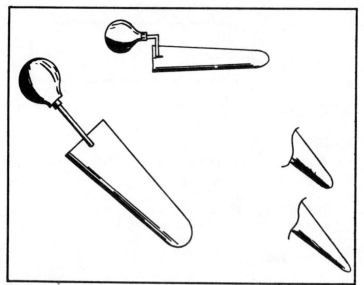

Fig. 4-1. Molder's trowels.

Heart Trowel

The heart trowel is a handy little trowel for general molding and as its name implies has a heart shaped blade. They run in size from 2 inches wide to 3 inches wide in ¼ inch steps (Fig. 4-2).

Core Maker's Trowel

The core maker's trowel is exactly like a finishing trowel with the exception that the blade is parallel to its entire length, and the nose is perfectly square. They come in widths of from 1 inch to 2 inches in ¼-inch steps and blade lengths of 4½ inches long to 7 inches long in ½-inch steps. This trowel is used to strike off core boxes and because it is parallel and square ended, a core can be easily trimmed or repaired and the sides can be squared up (Fig. 4-3).

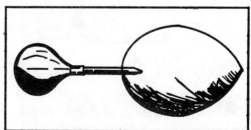

Fig. 4-2. Heart trowel.

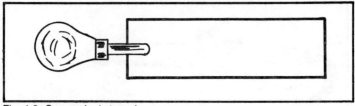

Fig. 4-3. Core maker's trowel.

BRUSHES AND SWABS

The most popular size brush for general bench work and light floor work is the block brush. It is 1⅛ inches wide and 9 inches long with four rows of bristles.

It's main use is to brush off the pattern, the matchplate, the bench and sometimes the molder himself.

Camel Hair Swab

A round camel hair brush (swab) is used to apply wet mold wash and core wash or blacking to a mold or core prior to drying. Most molders carry two sizes—one ⅝ inches in diameter and one ⅞ inches in diameter (Fig. 4-4).

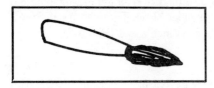

Fig. 4-4. Camel's hair swab.

Flax Swab

Also called a horse tail, the flax swab is used to swab the sand around the pattern at the junction of the pattern and sand. To dampen this sand to prevent it from breaking away when the pattern

Fig. 4-5. Use of flax swab.

Fig. 4-6. Molder's blow can.

is rapped and lifted from the sand, it is dampened by dipping it into a pail of water and shaking it out well. It is used on floor work where the pattern presents a fair-sized perimeter (Fig. 4-5).

The swab is also used by some molders to apply wet mold wash or blacking to a mold surface, however, this requires great dexterity.

Blow Can

The blow can is a simple mouth spray can used to apply liquid mold coats and washes to molds and cores, or with water to dampen a large area. It can also be operated with the air hose like the sucker (Fig. 4-6).

Bulb Sponge

The bulb sponge consists of a rubber bulb with a hollow brass stem terminating in a soft brush. The stem is pulled out and the bulb filled about three-fourths full of water. The stem is then replaced. The bulb sponge is used to swab around the pattern prior to drawing it—the same as the flax swab is used for floor molding. The bulb is gently squeezed to keep the brush wet while swabbing (Fig. 4-7).

SPRUES AND CUTTERS

You can turn sprues or purchase them in various diameters and use them to form sprue and pouring basins in sand molds. In time you will wind up with a wide selection (Fig. 4-8).

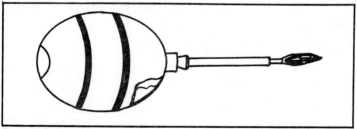

Fig. 4-7. Bulb sponge.

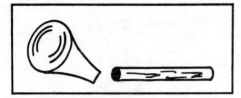

Fig. 4-8. Wood sprues.

Tubular Sprue Cutters

Tubular sprue cutters are tapered steel or brass tubes used to cut sprue holes in the cope half of a sand mold. They are sold in sizes from ⅞ to 1¼ inches in diameter. All are 6 inches long (Fig. 4-9).

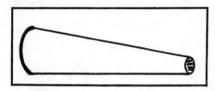

Fig. 4-9. Tubular sprue cutter.

Gate Cutter

The best gate cutter is made from a section of a Prince Albert Tobacco Can (Fig. 4-10).

To cut a gate or runner in green sand the tab is held between the thumb and first finger and the gate or runner is cut just as you would cut a groove or channel in wood with a gouge. The width is controlled by bending the cutter's sides in and out. The depth is controlled by the operator.

RAMMING AND RAPPING

Various tools are used by the metalcaster for the ramming and rapping operations involved during molding.

Bench Rammer

A bench rammer can be made of oiled maple. You can buy one or turn one on a lathe and band saw the peen end with its wedge shape.

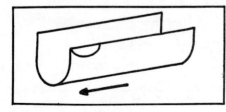

Fig. 4-10. Gate cutter.

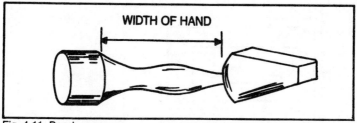

Fig. 4-11. Bench rammer.

The butt end is used to actually ram the sand into the flask around the pattern. The peen end is used to peen or ram the sand tightly around the inside perimeter of the flask to prevent the cope or drag mold from falling out when either half is moved, rolled over or lifted (Fig. 4-11).

Floor Rammer

The floor rammer has the same purpose and design as a bench rammer except the butt and peen ends are made of cast iron and attached to each end of a piece of pipe or hickory handle. The average length is 42 inches (Fig. 4-12).

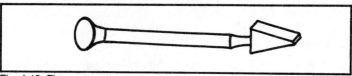

Fig. 4-12. Floor rammer.

Rapping Bar and Rapper

The rapping bar consists of a piece of brass or steel (cold roll) rod which is machined or ground to a tapered point. The rapper is made of steel or brass and is shaped exactly like the frame of a sling shot (Fig. 4-13).

The purpose of these tools is to rap or shake the pattern loose from the sand molder in order to draw it easily from the sand. What

Fig. 4-13. Rapping bar and rapper.

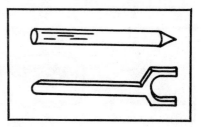

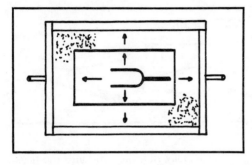

Fig. 4-14. Rapping.

you are doing is shaking the pattern in all directions which will drive the sand slightly away from the pattern. The resulting mold cavity will actually be a fuzz larger than the pattern. The pattern should only be rapped enough to free it from the sand. You will be able to see when it is loose all around by the movement of the pattern. Over rapping will distort the mold cavity which may or may not matter depending on how close you wish the casting to hold a tolerance (Fig. 4-14).

The operation is quite simple. The bar point of the rapping bar is pressed down into a dimple in the parting face of the pattern with the left hand. The rapper is used to strike the bar with the inner faces of the yoke.

Hand Riddles

You need two 18-inch diameter riddles—one with #4 mesh and one with #8 mesh galvanized iron screen. The #4 mesh will be used for general bench work and floor molding. It will riddle the first sand over the pattern before filling the mold with the shovel and ramming. The #8 mesh serves the same purpose but also applies fine facing sand to special jobs such as plaques or grave markers (Fig. 4-15).

Bellows

In general there are two types of bellows. One has a short snout and it is called a bench bellow. The other one has a long snout

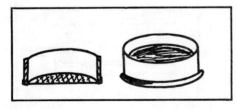

Fig. 4-15. Hand riddler.

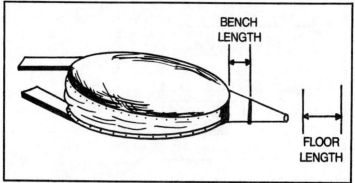

Fig. 4-16. Molder's bellows.

and is called a floor bellow. The theory is that a molder can wreck a bench mold when blowing it out with a long snout floor bellow by hitting the sand with the snout. This is true due to the different stance and angle when blowing a mold bench high, or on the ground. With care one only needs a 9 inch or 10 inch floor bellows to blow out cope and drag molds, sprue hole and gates (Fig. 4-16).

Strike Off Bar

Each time a mold is rammed up the sand must be struck off level with the flask cope and drag. The bar simply consists of a metal or hardwood straightedge of sufficient length (Fig. 4-17).

Raw Hide Mallet

The mallet is a minimum weight of 21 ounces and is used to rap pattern or draw pike to loosen the pattern from the sand so it can be smoothly withdrawn from the mold without damage to the mold or pattern.

MISCELLANEOUS TOOLS

There are several other tools necessary for the molder to successfully complete his metal casting operations.

Fig. 4-17. Strike-off tool.

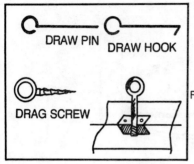

Fig. 4-18. Draw pins and draw plate.

Draw Pins, Screws and Hooks

These items are used to remove or draw the pattern from the mold. The draw pin is driven into the wooden pattern and used to lift it out as a handle. On a short small pattern one pin in the center will do it. If the pattern is long, use one on each end for a two-hand straight lift. The draw hook is used the same way. The draw screw is screwed into the pattern for a better purchase on heavier patterns, and prevents the pattern from accidently coming loose prematurely and falling, damaging the mold, pattern or both. In large patterns, plates are let into the pattern at its parting face which have a tapped hole into which a draw pin with a matching thread is screwed for lifting by hand or with a sling from a crane. Two or more are usually used (Fig. 4-18).

Loose metal patterns are drilled and tapped to receive a draw pin.

Sucker

The sucker is not a purchased tool but one made by the molder. It consists of two pieces of tubing and a tee of copper or iron (Fig. 4-19).

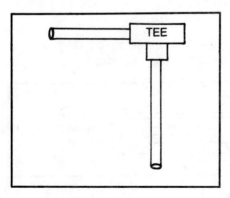

Fig. 4-19. Sucker.

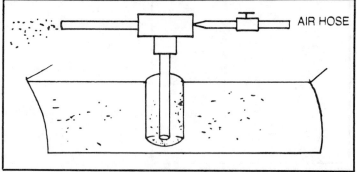

Fig. 4-20. Operation of sucker.

It's use is to clean out deep pockets in molds where the bellows and lifter fail or if the pocket or slot is just too dirty to spend the time a lifter would take or where it would be hard to see what you are doing with a lifter. The operation is very simple. You simply blow through the elbow with an air hose. This creates a vacuum in the long length which gives you in effect a vacuum cleaner with a long skinny snout. Now stick the long end down to where the problem is and blow through the elbow. These jobs will lift out small steel shot, a match stick or material which cannot be wetted such as parting powder or silica sand (Fig. 4-20).

A word of warning—watch in which direction you have the discharge end of your sucker pointed when sucking out a mold. You could accidently blow sand in someone's eyes.

The Bench Lifter

The bench lifter is a simple steel tool with a right-angled square foot on one end of a flat bent blade (Fig. 4-21).

This tool's biggest use is to repair sand molds and lift out any tramp or loose sand that might have fallen into a pocket (Fig. 4-22).

To remove dirt from a pocket that will not blow out with the bellows, you simply spit on the heel of the lifter and go down and pick it up. Then wipe off the heel and go back and slick the spot down a bit.

Fig. 4-21. Bench lifter.

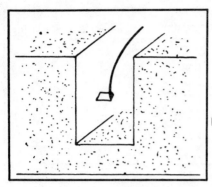

Fig. 4-22. Using the bench lifter.

Lifters come in all sizes from a bench lifter ¼ inch wide by 6 inches long to floor lifters 1 inch wide and 20 inches long. All have the same use. Many molders make their own lifters to suit the class of work then generally do.

Slick and Oval Spoon

The slick and oval spoon is a must for all molders. Again the size needed is determined by the work involved. Most molders have at least four sizes from one ¼ inch wide to one 2 inches wide. This tool is called a double ender in the trade. One end is a slick similar to a heart trowel blade but more oval shaped. The opposite end is spoon shaped. The outside or working surface is convex like the back of a spoon. Its inner face is concave. This face is never used and therefore is usually not finished smooth. When new it is painted black. Both faces of the slick blade are highly polished (Fig. 4-23).

The double ender is a general use molding tool used for slicking flat or concave surfaces, for opening-up sprues, etc.

Dust Bags

Two dust bags are needed by the molder—one 5 inches × 9 inches for parting powder and one 7 inches × 11 inches for graphite. These bags can be purchased from supply houses, but most give them away. You can, as I do, use an old sock (with no holes) for both parting and blacking. The parting bag is used to shake parting on the pattern and parting line of the mold faces. The blacking bag is used to dust blacking on a mold surface prior to sleeking.

Vent Wire

The vent wire is simply a slender pointed wire with a loop at its top used to punch vent holes in the cope and drag of a sand mold. It

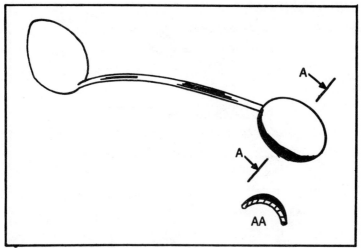
Fig. 4-23. Slick and oval.

provides easy access of steam and gases to the outside during the pouring of the casting. The venting is done prior to the pattern removal. The vent wire is pushed down into the sand to within close proximity of the pattern. The first and second finger straddles the wire each time it is withdrawn as a guide (Fig. 4-24).

FLASKS

Sand molds used in metal casting are made in wooden frames called *flasks*. They have no resemblance at all to the glass or metal liquid container most people think of when they see or hear the word flask.

They actually are open wooden frames that can be held together with pins and guides. They are separated during the mold preparation process and placed back together for pouring the mold without losing the original register. A flask with good pins and guides will close back together in the exact same place every time. This is essential to avoid shifts in the mold cavity with a resulting defect.

Fig. 4-24. Vent wire.

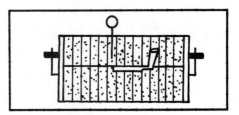

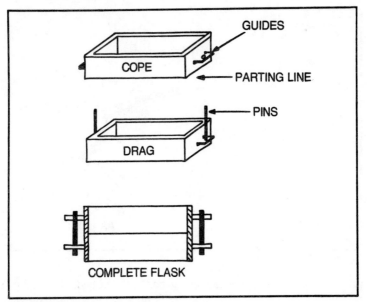

Fig. 4-25. Basic flask assembly.

Basic Flask

The top frame of the flask is called the *cope* and the bottom frame is called the *drag* (Fig. 4-25).

In some molding operations you need one or more sections between the cope and the drag. These sections are called *cheeks* (Fig. 4-26).

The pins and guides that are used to hold the sections together can be purchased in a wide variety of types and configurations—round and half round, double round and vee shaped. Single, double or triple vee shapes together with matching guides. Both pins and guides have attached mounting plates by which they can be bolted, screwed or welded to the halves which make up a complete flask set (Fig. 4-27).

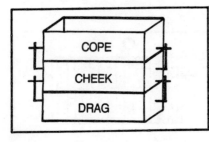

Fig. 4-26. Flask with cheek.

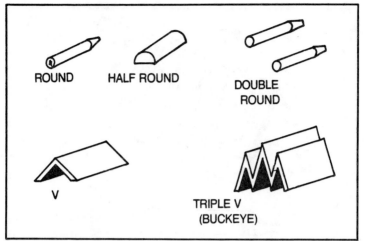

Fig. 4-27. Alignment pin types.

Home Made Wood Flask

In hobby or small shops you will find wood flasks with simple wood guides and pins. Although they work for a while they soon loosen or burn up from spilled metal. Only resort to this type for a one-time quick job for which you must build a quick flask to fit it (Fig. 4-28).

Floor Flasks

These are some flasks that are too large to be handled on the bench. These can be of wood or metal. The wood flasks are constructed in such a manner that the long sides provide the lifting handles for two-man lifting and handling.

With floor flasks, unlike bench flasks, when you reach a size of 18 inches × 18 inches to 30 inches × 30 inches, members are put in

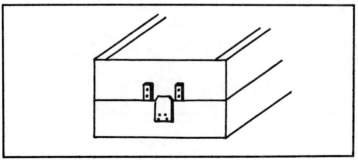

Fig. 4-28. Wood pins and guides.

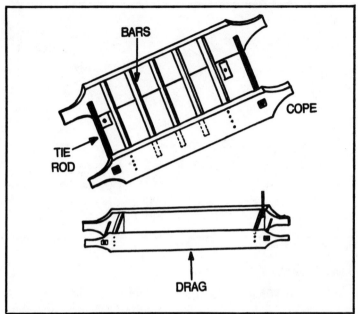

Fig. 4-29. Wood floor flask.

the cope which are called *bars*. These bars help support the weight of the sand in the cope to prevent it from dropping out. Bars are required in the drag half of the flask only when it is necessary to roll the job over and lift the drag instead of the cope in such cases they are called *grids* (Fig. 4-29).

The bars do not come all the way to the parting but clear the parting and portion of the pattern that is in the cope by an minimum of ½ inch. These bars in many cases have to be contoured to conform to the portion of the pattern that is in the cope (Fig. 4-30).

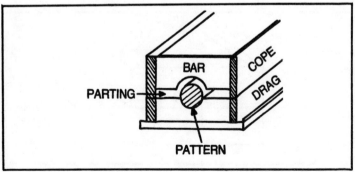

Fig. 4-30. Contoured cope bar.

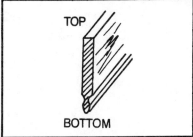

Fig. 4-31. Contoured bar.

In all cases the bars are brought to a dull point along their lower edge to make it possible to tuck and ram the sand firmly under the bars (Fig. 4-31).

The inside surfaces of both the flask and the bars in the cope section are often covered with large headed roofing nails with the head projecting ⅛ to ¼ inch. This gives the entire inner surface an excellent tooth and a good purchase on the sand.

Of course the floor flask like all others must be provided with suitable pins and guides on both ends.

Small floor flasks up to 30 inches × 30 inches with a cope and drag depth of from 5 to 8 inches can be made of 1-inch lumber. From 32 inches × 32 inches to 48 inches × 48 inches with 5-inch to 10-inch cope and drag, use 2-inch lumber. From 50 inches × 50 inches to 62 inches × 63 inches with 6-inch to 19-inch cope and drag, use 2½ inch lumber. From there up to a flask that measures a maximum of 85 inches × 85 inches with a cope and drag depth of 7 to 30 inches, use 3-inch lumber. The bars should be made of the same thickness of lumber as the flask. The number of bars required is generally determined so that you have a maximum of 6 inches of sand between them.

If your flask is extremely wide, you might need a cross bar between the bars. This bar is called a *chuck* (Fig. 4-32).

When you get to a flask larger than 85 × 85 inches you should move to welded steel.

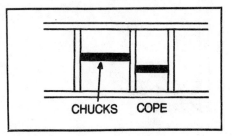

Fig. 4-32. Cope chucks.

CHUCKS   COPE

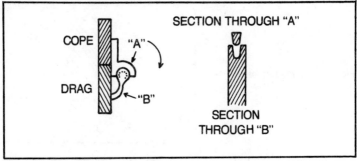

Fig. 4-33. Roll off hinge and guide.

Where the type of work permits, the floor flask can be fitted with roll off hinges. These hinges allow the flask to be opened like a book. The pattern is removed and the mold is closed. This operation is completed without lifting the cope (Fig. 4-33).

Large Steel Floor Flasks

Large steel floor flasks are more often equipped with female guides on both cope and drag halves with loose pins used for molding and closing. Some have single hole guides, some have double hole guides (Fig. 4-34).

Steel flasks can be purchased in all sizes with any size cope and drag depth, or any combination of different cope and drag depths and any type of pin and guide arrangement you desire.

Snap Flask

A snap flask is usually made of cherrywood. After the mold is made, the flask can be removed by opening the flask and lifting it off

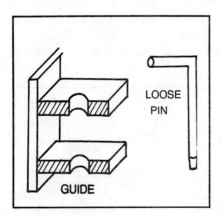

Fig. 4-34. Loose pin and guide.

of the mold, leaving the mold as a block of sand on the bottom board (Fig. 4-35).

In corner "A," both the cope and drag have a hinge and in corner "B," a cam locking device. In operation the cope and the drag locks are closed tight and the mold is made in the usual manner. When finished, the locks are opened and the flask is opened and removed from the mold. A typical snap flask hinge and lock is shown in Fig. 4- 36.

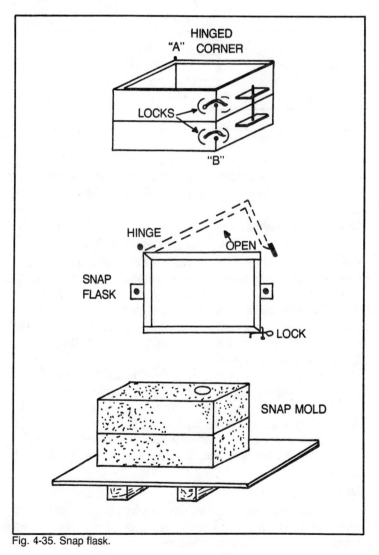

Fig. 4-35. Snap flask.

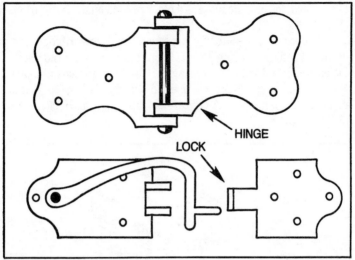

Fig. 4-36. Snap hardware.

The big advantage of the snap flask is that you need only one flask to make as many molds a day as you wish. With rigid flasks you need as many flasks as the number of molds you wish to put up at a time.

Other Types of Flasks

There are two other types of flasks that are removed from the mold. One is called a *tapered slip flask*. In this type of flask there is a strip of metal on the parting face of the cope which prevents the cope from dropping out while molding, which it would do without the strip, due to the taper on the cope's inner surface. After the mold is made and closed, this strip is retracted by a cam lever. When the strip is retracted the tapered flask is easily removed by simply lifting it from the mold (Fig. 4-37).

The other type is called a *pop-off flask*. In this type of flask, both corners that cope and drag diagonally across from one another are clamped by a cam locking device. When the mold is completed the cam is released and the spring loaded corners pop apart just enough to release the sand mold. These are quite popular and can be had in enormous sizes for floor and crane work as well as small bench sizes.

Jacket

A jacket is a wood or metal frame which is placed around a mold made in a snap flask during pouring to support the mold and prevent a run out between the cope and drag.

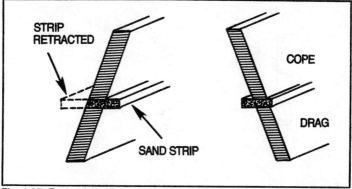

Fig. 4-37. Tapered slip flask.

The common practice is to have as many jackets for each size snap mold as can be poured at one time. When the molds have solidified sufficiently the jackets are removed and placed on the next set of molds to be poured. These are called *jumping jackets*. Jackets must be carefully placed on the molds so as not to shift the cope and drag. The jackets must be kept in good shape and fit the molds like a glove. A tapered snap flask and tapered jackets work best due to the taper (Fig. 4-38).

Upset

When you need a taller cope section or drag section than the flask on hand, use a frame of the required additional depth needed. The inside of the frame is provided with four or more metal strips or tabs which fit down into the flask to hold it in place during molding. These strips or bands (they may be sheet iron bands) are called *upsets*. The term *to upset* a cope or drag means to add depth to it (Fig. 4-39).

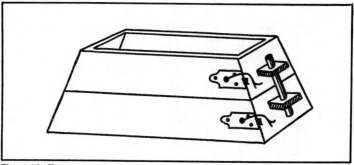

Fig. 4-38. Tapered snap flask.

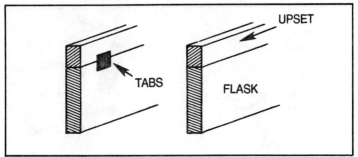

Fig. 4-39. Upset.

BOARDS

All boards for molding should be smooth. The pattern and flask rest on such boards when you start to make the mold.

Molding Board

The molding board is a smooth board on which to rest the pattern and flask when starting to make a mold. The board should be as large as the outside of the flask and stiff enough to support the sand and pattern without springing when the sand is rammed. One is needed for each size of flask. Suitable cleats are nailed to the underside of the board. Their purpose is to stiffen the board and to raise it from the bench or floor to allow you to get your fingers under the board to roll over the mold (Fig. 4-40).

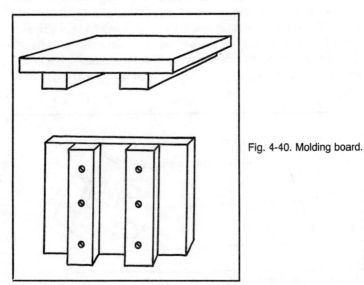

Fig. 4-40. Molding board.

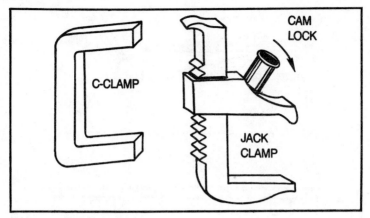

Fig. 4-41. Floor mold clamps.

Bottom Board

The bottom board is the board used to support the sand mold until the mold is poured. It is constructed like the molding board, but it need not be smooth, only level and stiff. Bottom boards from 10 inches × 16 inches to 18 inches × 30 inches should be made of 1-inch thick lumber with two cleats made of 2 × 3-inch stock. Bottom boards 48 × 30 inches use 1½-inch thick lumber with three

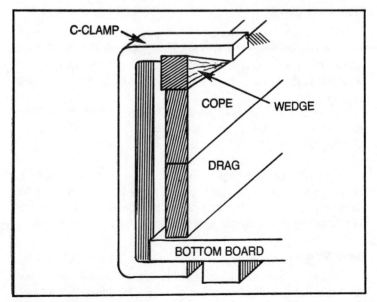

Fig. 4-42. The use of C-clamps.

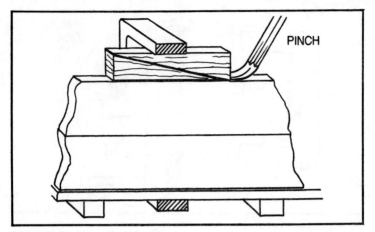

Fig. 4-43. Double wedged C-clamps.

cleats made of 3 × 3-inch stock. Bottom boards 50 to 80 inches use 2-inch thick lumber with four to five cleats made of 3 × 4-inch stock.

CLAMPS AND WEIGHTS

Floor molds are always clamped to pour. Sometimes they are weighted, sometimes not, but they are always clamped.

Floor Mold Clamps

There are two basic types of clamps used (Fig. 4-41).

The C-clamp is cast iron or square steel stock. In operation the bottom foot is placed under the bottom board and a wooden or steel wedge is tapped between the top foot and the top of the flask side (Fig. 4-42).

A better practice is to place two wedges between the clamp foot and flask by hand, then tighten them with a small pinch bar. Driving them too tightly jolts the mold and could cause internal damage such as a drop (Fig. 4-43).

The jack clamp is the preferred, but most expensive clamp. Its operation is simple. The foot is placed under the bottom board and the clamping foot is slid down against the cope flask. A bar is placed in the cam lever and pushed down until the clamp is snug and tight.

Mold Weights

All snap flask work is weighted before molding. In most cases all that is needed is a standard snap weight. It is usually cast to suit your own snap sizes (Fig. 4-44).

82

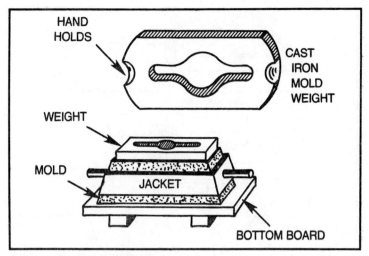

Fig. 4-44. Typical mold weight.

The snap flask mold weights are from 1½ to 2 inches thick cast iron and weigh from 35 to 50 pounds for a 12 × 16-inch mold. The rounded cross-shaped opening through the weight is to accommodate pouring the metal into the mold. The weights are always set so that the pouring basin is free to easily see and pour, not too close to the weight.

In some cases two or more weights are used per mold and are stacked.

Another type of mold weight which I favor for snap and small rigid flask bench work is the sad iron type of weight. They look like the old-time irons heated on the stove for pressing clothes, only heavier (Fig. 4-45).

Another method used to weight bench molds in rigid full flasks, whether wood or metal, is to stack the molds three high, similar to steps. This form puts weight only on the top mold. This practice

Fig. 4-45. Sad iron weight.

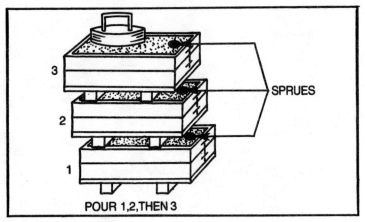

Fig. 4-46. Stacked molds.

was quite common in a great many foundries and is still used today in home casting shops (Fig. 4-46).

The lifting or buoyant force on the cope is the product of the horizontal area of the cavity in the cope, the height of the head of metal above this area and the density of the metal. This lifting force on the cope has to be dealt with by the metalcaster. If the force is greater than the weight of the cope, the cope will be lifted by this force when the mold is poured. It will run out at the joint resulting in a lost casting. In order to determine how much weight has to be placed on the cope to prevent a lift of the cope, multiply the area of the surface of the metal pressing against the cope by the depth of the cope above the casting and the product by 0.26 for iron, 0.30 for brass and 0.09 for aluminum. These figures represent the weight in pounds of 1 cubic inch of iron, brass or aluminum, respectively.

If you are casting a flat plate 12 × 12 inches, the surface area against the cope will be 144 square inches. If this would be molded in a flask which measured 18 × 18 inches with a 5-inch cope and the casting metal would be red brass, we'd have 144 × 5 inches (the height of the cope × 0.30 (the weight of 1 cubic inch of brass). You'd wind up with a lifting force of 216 pounds. One cubic inch of rammed molding sand weighs 0.06 pounds. With that factor, the cope weight would be 18 × 18 × 5 inches × 0.06 or 97.2 pounds. If you subtract the cope weight from the lifting force—216 pounds minus 97.2 pounds—you are still short by about 119 pounds. This would be the weight you'd have to add to the cope to prevent it from lifting, with no safety factor. With a 20 percent safety factor, add 24 more pounds. You should put 150 pounds on the cope.

Chapter 5
Metal Melting Devices

Nonferrous metals are melted in numerous types of furnaces. The crucible furnace is the most common, and lowest in initial cost.

CRUCIBLE FURNACES

The furnace is essentially a refractory-lined cylinder with a refractory cover, equipped with a burner and blower (Fig. 5-1) for the intense combustion of oil or gas. The metal is melted in a crucible (pot) made of clay and graphite, or silica carbide, which is placed in the furnace. When the melting is complete, the furnace is turned off, the furnace cover is opened and the crucible removed with tongs and placed in a pouring shank. Then the liquid metal is poured into prepared molds.

A crucible furnace can be made from an old metal drum or metal garbage can, a few pipe fittings, the blower from an old vacuum cleaner and 30 or 40 dollars worth of castable refractory material. The refractory suppliers can also furnish you with complete advice on what to buy for your particular purpose, along with tips on how to handle the material. Crucibles come in sizes from as small as your thumb to a number 400, which will hold 1200 pounds of molten bronze.

Homemade melting furnaces are simple to construct, and are fun to build and operate. A furnace for a No. 20 crucible, which has a single pour capacity of 60 pounds for bronze, or 20 pounds for aluminum, would be considered average in size.

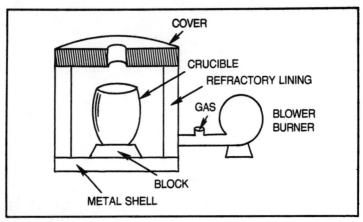

Fig. 5-1. Crucible furnace.

By weight, crucibles will hold three times as much bronze as aluminum. Always allow some clearance below the top of the crucible for safety—it is dangerous to melt a brimming potful.

Let's assume, for this discussion, that you have chosen a No. 20 crucible: 10⅛ inches high, 7 13/16 inches across the top and bottom and 8½ inches at its bilge, or widest point (its shape resembles that of a barrel). This crucible requires a base in the form of a truncated cone 6 inches across the top, 7 inches across the bottom and 5 inches high. Now we have 15 inches of height and 8 inches in diameter to go into the furnace.

There must be sufficient distance between the crucible and the furnace lining for correct combustion, and for room to fit open tongs around the crucible. Also, enough space between the furnace bottom and the covers is needed for correct combustion and exhaust through the cover opening. A good general rule is to allow 2½ to 3 inches of clearance between the furnace wall and the bilge. Leave 3 inches between the top of the crucible and the cover and above the furnace bottom. In general, the lining should be a minimum of 4 inches thick to insure good insulation.

With this we need a shell for our furnace 22½ inches (inside diameter) by 22⅛ inches high with a cover band (Fig. 5-2) 22½ inches (inside diameter) by 4 inches in height. A safety hole directly in the front of the furnace, flush with the bottom and 3 inches square, is needed. Should the crucible break, the metal would run out of the furnace through the hole.

Without a safety hole, metal could run into the burner pipe, or simply fill up the bottom of the furnace and solidify there, leaving

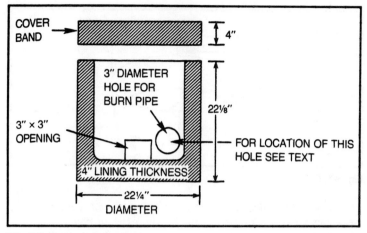

Fig. 5-2. Furnace dimensions.

you with a hunk of metal next to impossible to remove short of tearing everything apart.

The shell is completed by making a 3-inch hole 6 inches above the bottom of the shell, a third of the shell's circumference away from the safety hole. This is where the burner pipe enters. The burner pipe is brought in 2 inches above the refractory lining of the bottom, and off-center from the diameter of the shell so that the flame coming will circle the space between the crucible and the lining, spiral around the crucible and out of the vent hole; this gives the highest and most even heat (Fig. 5-3.)

The cover consists of a metal band formed into a ring 22½ inches in diameter, but tapered slightly (Fig. 5-4). Tabs or lifting ears must be riveted on directly across from each other on the cover "ring." They should be drilled with holes to clear a ½-inch pipe which will be used to remove the cover.

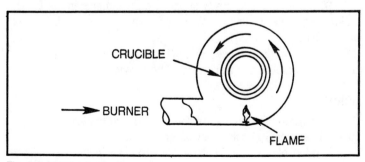

Fig. 5-3. Heating method.

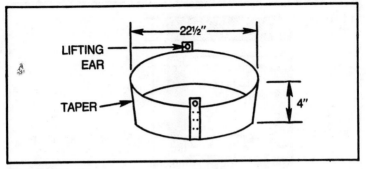

Fig. 5-4. Cover ring.

The cover ring is placed on a smooth floor covered with newspaper in preparation for making the exhaust hole.

Within the center of the cover ring, place a 6-inch tin can or jar as a form for the exhaust vent. The castable refractory is now mixed according to the manufacturer's directions, tamped firmly in place around the vent form, leveled off smoothly with the top of the ring and left to set overnight.

Now we are ready to line the furnace body. Place a heavy cardboard sleeve 14½ inches in diameter and long enough to extend slightly above the top of the shell in the center of the furnace—22 or 23 inches tall should do.

Once the sleeve is centered, fill the inside of the sleeve with dry sand to give it added strength. Hold it in place while tamping in the lining all the way to the top. Two plugs will be needed for the furnace shell while the lining is being installed: one to fit through the safety hole and against the cardboard sleeve and another to fit the burner port. Dimensions for the plugs are given in Fig. 5-5.

Coat each plug with heavy oil or grease, and fit both into their respective places. Use small wooden wedges to get a snug fit.

With both plugs securely in place, start tamping in the lining, making sure the castable refractory is well compressed. Do not place too much material between the sleeve and the shell at a time; doing so will produce spaces. When the top of the shell is reached, trowel and smooth the refractory. The furnace is now complete.

BURNERS

Now you must have a suitable burner and blower—a blower that will deliver 300 cubic feet of air per minute.

A good blower, particularly one from an industrial vacuum cleaner, will deliver enough air at the right pressure.

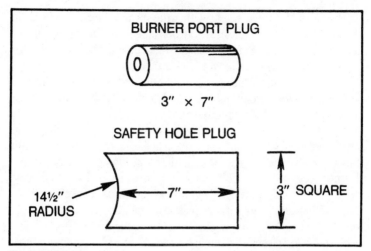

Fig. 5-5. Hole plugs.

The simplest type of gas burner can be made from a pipe 2 feet long. Heat the pipe red-hot at one end and hammer the end partially closed—not much over ¼ inch in all around—so that you are left with an opening 1½ inches wide. Heat the pipe again, 1 foot from the hammered end. Hammer in a neck with an inside diameter 1½ inches that is 1 inch wide (Fig. 5-6). In the bottom of the groove left from this *necking down,* burn a hole and fit a ½-inch pipe snugly into it. Weld it in place. You have just made a nipple for the gas line.

Insert the burner pipe into the burner port in the furnace body, attach the blower to the opposite end, connect a stopcock to the nipple and connect the whole thing to a gas outlet.

The intake of the blower must be provided with a shutter or damper to regulate air going to the burner.

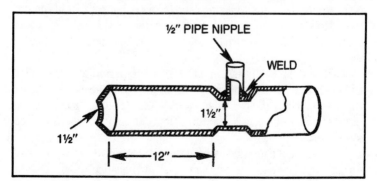

Fig. 5-6. Gas burner.

A butterfly valve will fit within the burner pipe. Whether you decide to control the air in the burner pipe or at the blower intake doesn't matter; but it must be constructed in a manner that provides positive action without moving of its own accord.

The newly-lined furnace should be slowly dried by building a wood fire inside and letting the wood burn down to coals. It takes about two days of this treatment to be safe.

For the first heat, put two thicknesses of cardboard on the crucible support block and place the crucible on the cardboard. With the furnace cover off, place the metal charge loosely in the crucible. Do not wedge the metal in because it must have room to expand without restriction. Wedging metal in a crucible can cause it to split.

Now place a wad of gunnysack material dampened with fuel oil, or charcoal lighting fuel, about 1 foot from the burner port, in line with the firing direction. The lighter wad should be jammed snugly between the support block and the furnace wall. This will prevent it from being blown out of the furnace or away from the burner. The wad has to remain in place, burning until the furnace wall reaches ignition temperature. Prepare to fire the burner by opening the blower's air control valve halfway. Light the wad and allow it to burn briskly. The blower is started up. Check to see if the wad is still burning briskly. Should it blow out, turn off the blower and start over. Once you have determined that the lighter wad will burn with the blower on, open the gas valve until you get ignition.

Adjust the gas to the point that produces maximum ignition—the loudest roar in the furnace. At this point, you have maximum combustion for the blower's output. Allow the furnace to run for 5 minutes at this setting. After 5 minutes have elapsed, the furnace wall should be hot enough to maintain combustion. Place the cover on the furnace, and advance gas intake and blower output to the point where the gas is wide open and the air is adjusted for maximum roar. Now advance the blower output slightly. This will give you a slight oxidation in the furnace which is the best condition for melting.

During the lighting up and the 5-minute period with the cover off, keep your hand on the gas valve. Should you lose ignition during this period, close the gas valve at once, let the blower run a minute or two to clear the gas-air mixture, then close down the blower and start over.

Never light a furnace without the blower delivering at least half its capacity. A too low setting can result in a backfire into the blower due to back pressure in the furnace. The blower has to be blowing

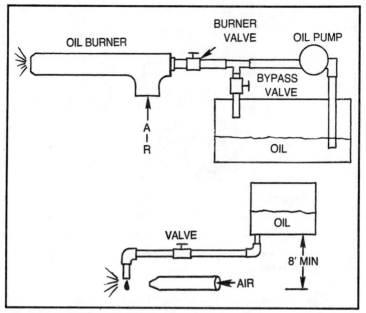

Fig. 5-7. Oil burner designs.

strong enough to overcome back pressure so that the ignition takes place in the furnace. Do not light a furnace with the cover on for the same reason. Should the power fail, and the blower stop, immediately turn off the gas. Never leave a furnace unattended during the melting process. To shut down the furnace, close off the gas first, then the blower. Oil-fired furnaces (Fig. 5-7) are lighted and shut down the same way.

COKE FIRED FURNACE

Although there are not many in use today, because they are slow and messy in operation, natural-draft coke-fired furnaces are simple in construction and have a low initial cost.

To operate a coke-fired furnace (Fig. 5-8), place a fire brick on the grate, cover the grate with wood and paper kindling and cover the kindling with a 3-inch layer of coke. Light the kindling and when the coke ignites and glows red, add another layer. Add layers until you have a deep bed of red-hot coke. Remove some coke from the center and place the crucible in this well; then build up a layer of fresh coke to the top of the crucible. Fill the crucible with metal and put the cover on. Adjust the draft to promote maximum combustion. As the coke burns down and drops into the ash pit, replace it.

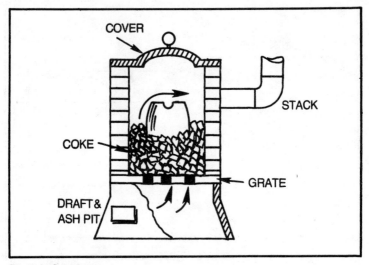

Fig. 5-8. Coke burner.

When the metal has reached the desired temperature, pick out coke from around the crucible, and pull out the crucible with the tongs; place it in the shank and pour. The brick in the bottom on the grate is used to support the crucible should it work its way down too far in the furnace.

A crucible furnace can also be easily constructed in the ground (Fig. 5-9) in areas where a water table is not too close to the surface.

TAPPED CRUCIBLE FURNACE

The tap-type crucible furnace is commonly used by people who work alone, casting small to medium-sized work, or who have a minimum of headroom to pull out the crucible. The furnace (Fig. 5-10) is built on four legs with a front opening at the bottom. This opening should be at least 4 inches by 4 inches. A hole is drilled in the front of the crucible, level with the bottom, to allow the metal to run out.

The furnace is lit and charged in the conventional manner. After 5 minutes the brick is removed, the cover is closed and combustion is adjusted. A refractory-lined ladle is preheated by placing it on bricks above the exhaust hole in the cover during the melting.

When the metal is ready to pour, the hot ladle is removed with a hand shank (handle) and placed in front of the furnace.

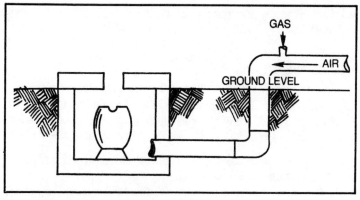

Fig. 5-9. Pit furnace.

A clay plug in the crucible's tap hole is picked out with a tapping bar made from a ½-inch steel bar sharpened to a point. The metal runs into the hand ladle. The ladle is then used to pour the mold.

To close up the tap hole for the next heating, a cone of bod (plus) mix is formed on a bod rod, and firmly pushed into the hole. The bod rod is given a little twist to release the bod. The heat of the furnace and crucible bakes the bod into place.

The crucible is placed in the furnace with the tap hole facing the opening in the front of the furnace. A refractory trough is made from the bottom of the tap hole to the furnace opening to carry the tapped metal out to the ladle. The operation is very simple. The tap hole is stopped with a bod made of one part sharp silica sand to one part milled fireclay. The mixture is dampened to form stiff mud and pressed firmly into the hole. To light the furnace, the lid is simply propped up at its edge with a chunk of firebrick. If the hole should

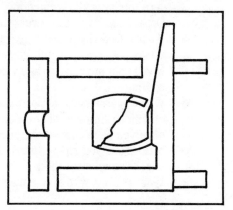

Fig. 5-10. Tap type furnace.

weep or leak at any time during heating (very rare), put a ball of bod mix on the bod rod and press it over the leak.

POURING THE CASTING MOLD

The actual pouring of molten metal into molds is a very important phase of the casting operation. More casting are lost due to faulty pouring than to any other single cause. Some basic rules for *gravity casting* a mold poured from a crucible or ladle are:

- Pour with the lip of the ladle or crucible as close to the pouring basin as possible.
- Keep the pouring basin full (choked) during the entire pour.
- Keep the pouring lip clean to avoid dirt or a double stream.
- Use slightly more metal than you think you'll need.
- Pour on the hot side—more castings are lost by pouring too cold, rather than too hot.
- Once a choke is started, do not reduce the stream of metal.
- Do not dribble metal into the mold or interrupt the stream of metal.
- If a mold cracks and the metal starts to run out, don't try to save it.
- If a mold starts to spit metal from the pouring basic or vents, stop pouring. Continuing to fill a wet mold that is spitting back can result in a bomb.
- Don't use weak or faulty tongs, or shanks.
- Keep the pouring area clean and allow plenty of room for sure footing and maneuvering.
- Do not pour with thin, weak crucibles.
- Wear a face shield and leggings.
- When pouring at night, dust the pouring basin with wheat flour in a bag to make a more visible target for the pouring.
- When pouring with a two-man shank, make sure the other man knows what he is doing (not move or jerk back once you have started the pour).
- Make sure you have a good, dry pig bed to hold any excess metal after the pour.
- When pouring with a hand shank, rest the shank on your knee.

- When pouring several molds in a row with a hand shank, start at one end and back up as you go. Going forward to pour brings the knuckles of the hand closest to the ladle over the mold just poured.
- When pouring several molds from a single ladle or crucible, pour light, thin castings first (the metal is getting colder by the minute).
- Don't try to pour too many molds at a single heating.
- Make sure that flasks are closed and clamped or weighted properly.
- Don't pour when you are in an awkward position. You must be relaxed.
- Don't try to lift and pour too much metal by hand. Use a crane or chain fall on a job boom. More than 40 pounds in a hand shank, or 200 pounds in a two-man bull shank begs disaster.
- If the metal in the ladle or crucible is not bright, clean, clear and hot, don't pour it.

Large molds are often poured by placing a sheet of asbestos over the sprue, on top of which is a large molten metal reservoir made of sand. The amount of metal required for the casting is poured into the reservoir, and a rod is used to puncture the asbestos sheet (Fig. 5-11), allowing the metal to fill the mold.

KILNS

The types of kilns used for metal casting are so numerous that a large volume would be needed on this subject alone. The factors

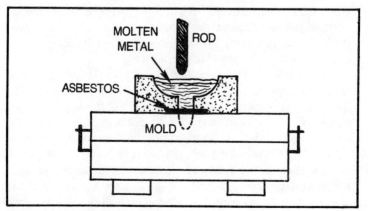

Fig. 5-11. Pouring reservoir.

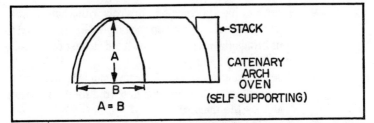

Fig. 5-12. The catenary-arch oven. The shape of the arch can be duplicated by suspending a flexible core with a uniform cross section between two points.

influencing their diveristy are mold size, mold material, pouring technique and the availability of fuels.

Small, tin-can molds and stainless-steel flasks for plaster-bonded investments require only a small, box-type oven, fired with a simple gas burner because the maximum temperature won't exceed 1250°F.

An electric kiln would be just suitable for the same reason. These can be easily built or purchased cheaply.

The catenary arch, downdraft oven (Fig. 5-12) is recommended for larger flask molds and hand-built molds.

Large molds to be poured in place can be calcined in a stacked kiln built on the spot around the mold from a diagonally cut oil drum lined with industrial insulation (Fig. 5-13). Regardless of the size chosen or the type of construction, downdraft kilns are best for many molds, especially lost wax molds because the heat is drawn downward and under the molds, promoting even firing and firing temperatures.

The refractories used for the kiln, regardless of its style or size, are largely a matter of choice. However, with today's modern castable refractories, one-piece construction is by far the easiest method of construction. Brick requires considerably more skill in its use, and is usually heavier in overall weight.

Suppliers of refractory materials carry a wide variety of castable refractories and cast shapes. They can advise you about which is best for your particular needs. Regardless of the kind of kiln you choose, it must be designed to *dry* the molds, not to simply bake or cook them. Dry air must be taken into the oven and take moisture with it when it leaves the stack. If a kiln does not do this, the air will become stagnant and saturated with moisture. When this happens, the mold stops drying and just bakes like a roast in a covered pot. Venting must be good enough to bring in fresh air continuously as moisture-laden air exits.

CUPOLAS

Although most investment materials are limited to metals poured at 2000°F or less, ceramic shell molds, and any of the sand systems, are suitable for ferrous castings. Due to a relatively low shrinkage factor and the superior fluidity of cast iron, you need only a minimum amount of gating and risers. The largest percentage of cast iron produced is melted in the cupola (Fig. 5-14). It is basically a miniature blast furnace. The cupola shown is the basic furnace used to reduce copper ore to *matte* copper. The cupola, besides reducing ore, can melt bronze, brass or cast iron. Although the cupola is usually thought of as a cast-iron melting furnace, it is excellent for melting bronze, in particular, silicon bronze and bronze low in lead or zinc content. (I have used a small, homemade cupola for many years for all my melting.)

Because the cupola melts continuously—as long as it is stoked and fed, it will melt charge after charge. The big advantage of the cupola is that you can pour a large quantity of molds, and collect a sufficient number of charges in a big ladle to pour castings in just about any weight you wish. You can also melt single batches of cast iron or bronze. Commercial cupolas range in size from a No. 0 with an inside diameter of 18 inches that will melt a ton of iron in an hour, to a No. 12, 84 inches in diameter that will melt more than 33 tons an hour.

A cupola is basically a refractory-lined cylinder with openings at the top for the escape of gases and the introduction of a charge. Smaller openings at the bottom are provided for the air blast and the release of molten metal and slag. A bed of fuel is laid on the cupola and ignited, after which alternate layers of metal and fuel are put

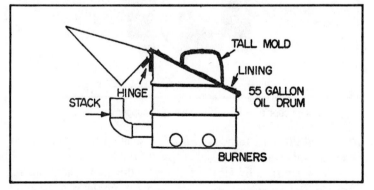

Fig. 5-13. A stacked kiln, built around a mold to be filled at the site of its construction.

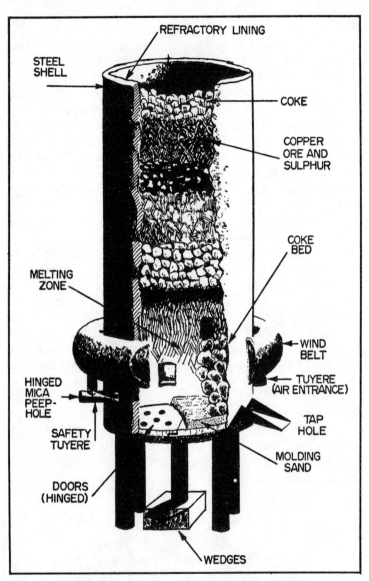

REFRACTORY LINING

STEEL SHELL

COKE

COPPER ORE AND SULPHUR

COKE BED

MELTING ZONE

WIND BELT

HINGED MICA PEEP-HOLE

TUYERE (AIR ENTRANCE)

TAP HOLE

SAFETY TUYERE

MOLDING SAND

DOORS (HINGED)

WEDGES

Fig. 5-14. The cupola used in the smelting of copper ore.

down and the blast turned on. If a few simple rules are followed, melting will begin quickly and continue for a long time.

The cupola is by far the simplest and most efficient melting furnace. Before getting into its actual operation, let's examine the cupola's construction from the bottom up.

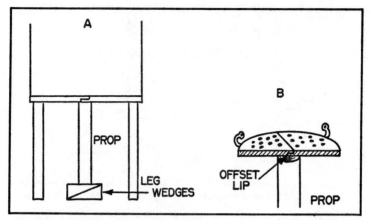

Fig. 5-15. The bottom of the cupola (A) is high enough off the floor to allow 6 inches of clearance for the open doors (B).

The cupola is supported on four legs and fitted with a pair of perforated, cast-iron doors underneath. The legs are long enough to allow the doors 6 inches of clearance above the floor when they are open (hanging down). When the doors are closed, they are held by a metal pipe wedged in place tightly with a pair of steel or cast-iron wedges (Fig. 5-15A). The doors have a mating offset lip (Fig. 5-15B) and ledge to insure a good fit. Four or more openings are cut through the lining and shell about 1 foot from the doors for tuyeres. They fan out at the inside of the lining and make an almost continuous opening by being close to other adjoining tuyeres (Fig. 5-16). The tuyeres are usually formed in cast-iron tuyere boxes installed during the lining of the cupola. To provide a blast of air through the tuyeres, the cupola is constructed with a *wind belt*—a sheetmetal tube encircling the outside of the cupola. There are two basic types. In the most common, air is introduced into the tuyeres through elbow connections extending from the bottom of the wind

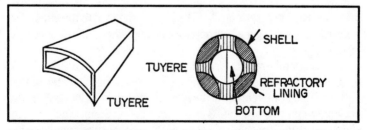

Fig. 5-16. The opening of the tuyeres inside the cupola are so close as to create practically one continuous opening.

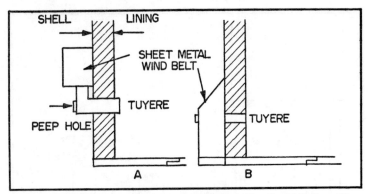

Fig. 5-17. A wind belt can be either above the tuyeres (A) or level with them (B).

belt and terminating at the outer tuyere opening (Fig. 5-17A). These are called *pendulum* tuyeres. In the second type, the wind belt encircles the tuyeres at their level (Fig. 5-17B).

It is customary to place one tuyere lower than the rest; this is a *safety* tuyere. Its purpose is to prevent molten metal from rising above the bottom of the tuyeres and spilling, burning up the wind belt or causing greater havoc. Should the metal rise too high it will spill over into the safety tuyere which is fitted with a lead disc covering a hole in the bottom of the pipe that supplies the blast. The disc would melt quickly, allowing the metal to drop through, thus preventing damage to the cupola.

A ladle is usually placed on the ground below the disc. Each tuyere has a peep hole fitted with a cover and mica window. The cover (Fig. 5-18) can be swung aside for access to the tuyeres, or locked in place when the cupola is under blast.

Blast air is supplied to the wind belt by a blast pipe. It enters the wind belt at a point between tuyeres. The blast circulates through the wind belt (Fig. 5-19) and enters the cupola through the tuyeres. This type of construction seems to be the best in that the blast is divided more or less evenly, delivering the same volume and pressure to all tuyeres equally. The blast is supplied to the cupola by a blower whose required output is based upon the inside dimensions of the cupola.

Besides tuyeres, the cupola must have a tap hole (Fig. 5-20) through which molten metal can be removed. This is located at the bottom front, between two tuyeres. A slag hole is located in the back between two tuyeres, 3 or 4 inches below them. The legs are sections of pipe attached to a plate used to anchor the cupola to the floor.

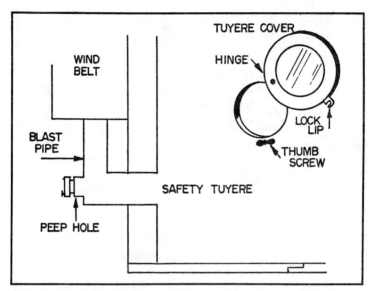

Fig. 5-18. Conditions within the cupola can be observed through a peep hole in the lowest of tuyere pipes.

A truly workable cupola, and one that will operate as safely and efficiently as possible, must be constructed according to certain design criteria. The refractory lining, for example, should never be less than 4½ inches thick. The diameter of the lining itself is determined by its thickness. For one thinner than 7 inches, the inside diameter should be no more than 36 inches—a thickness greater than that dictates a diameter from 41 to 56 inches.

Blower output should be 2.5 CFM (cubic feet per minute) per square inch of inside lining area. Always round off the output figure

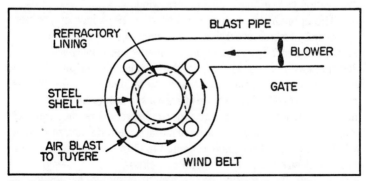

Fig. 5-19. The blast is divided evenly among the tuyeres by circulation around the wind belt.

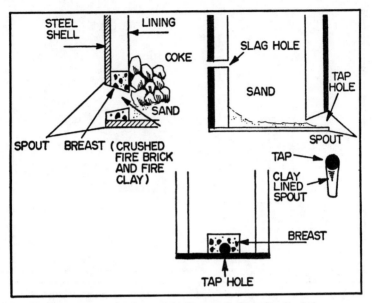

Fig. 5-20. Molten metal is drawn from the cupola through a tap hole.

you derive to the next highest blower size available, for safety. The level of the coke bed above the tuyeres is determined by blast pressure. Its height should be equal to the square root of the blast pressure times 10.5 plus 6.

A good way to determine the correct bed height is to note the length of time it takes the first drops of iron to fall past the tuyeres after turning on the blast. Iron should be seen at the tuyeres in 4 or 5 minutes; at the tap, in 6 to 8 minutes. If iron can be seen sooner, the bed of coke is too low; if seen later, the bed is too high. This observation can be made at a tuyere window.

The optimum melting rate is based on 10 pounds of iron per square inch of cupola area; that is, a 12-inch cupola should melt 1130 pounds per hour. The amount of iron needed per total charge—fuel plus metal—is based on a *melt ratio,* a value determined by the weight of the coke charges. Enough metal melted by the time 4 inches of the fuel bed have been burned away is optimum. A layer of fuel previously placed above the metal charge will replace the fuel consumed. When another charge of metal is melted away, more fuel will descend. The process is repeated until the end of the *heat,* or until the desired amount of metal is melted.

The constant relating to the weight of metal charges is the weight of 4 inches of coke. Usually, a metal ring 4 inches deep is made that has the same diameter inside as the cupola. It is filled with coke, and then the coke is weighed.

The melt ratio for iron is 9 to 1—9 pounds of iron for each pound of coke. If a cupola took 1 pound of coke to reach a 4-inch level, each iron charge would weigh 45 pounds and each coke charge would be 4 inches high. The melt ratio for copper or bronze is 20 to 1.

Some small cupolas (24 inches tall and 24 inches wide) are used for *batch melting* (one charge of metal).

These extremely short cupolas have a cover similar to those on furnaces used to melt bronze, and have their tuyeres close to the bottom. After each melt, the bed is built up and a new metal charge is added. This is done throughout the day, whenever a melt is needed. The cover has a 6-inch exhaust hole in its center. The cover provides back pressure to drive the blast downward to compensate for the cupola's height.

No two cupolas operate alike, but the guidelines given will put you in the ballpark. The most common type of homemade cupola is two 55-gallon oil drums welded end to end. They will do a remarkable job if built well and properly operated.

CUPOLA OPERATIONS

After raising the bottom doors and propping them up firmly, tamp down a compact layer of tempered molding sand on the bottom of the cupola, making it slope downward from the back and sides toward the tap hole. Cover the bottom with light, dry kindling and light it.

Add enough wood to get a good brisk fire going—approximately half of the bed coke. When this has burned through,

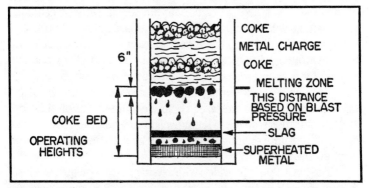

Fig. 5-21. The operation of a cupola properly charged with fuel and metal.

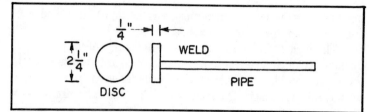

Fig. 5-22. A bod bar used to form and install a bod. A clay plug is used for the tap hole.

add the remaining half. When the entire bed coke is burning brightly (as seen through the top), add enough coke to bring it to the correct level. At this point, put in the first metal charge, followed by 4 inches of coke. Then add the next metal charge and so forth, until the desired number of charges (Fig. 5-21) are in the stack. Allow the cupola to "soak" for about a half-hour. (Whenever the blast is off, open one or more tuyeres to prevent the accumulation of explosive gases in the wind belt.)

The bed should now be properly *burned through* and the cupola charged with alternate layers of metal and coke. Close the tap hole with a bod (clay plug). The bod mix is fire clay and sand in equal amounts, tempered with enough water to make a pliable mixture. Make the bod on the end of a bod rod (Fig. 5-22) by first wetting the pad on the bod rod and forming a ball of bod mix there; then form it into a cone. Press the bod firmly into the tap hole, twist the bar to detach the body and slide the bar closed. Then turn on the blast and start the metal. The metal will collect in the bottom of the cupola with slag floating on top. When sufficient metal is in the well area, the slag will flow from the slag hole. When iron appears at the slag hole, tap the cupola by picking out the bod with a *tapping bar*—a pointed rod ½ to ¾ inch in diameter with a ring at one end. When enough metal has been tapped, plug the cupola with a fresh bod.

Melting will continue as long as the blast is maintained and metal and coke charges are added. Occasionally, chilled slag will build up around the tuyeres and impede the air blast. They must be cleared by punching them loose with a blunt bar (Fig. 5-23). The cupola can be held in a nonmelting state for several hours. Open the tap hole and drain the well area completely; then shut off the blast and open a tuyere cover. Melting will stop when the blast is shut down. To resume melting, close the tap hole, close the tuyere and start up the air blast.

The blast pipe is provided with a *blast gate* to regulate the blast. A blast gate is helpful to keep track of conditions in the cupola. As you may recall, the fuel bed level above the tuyeres is a function of blast pressure—low blast pressure for a low bed, high pressure for a high bed.

To achieve proper melting, the metal must be melted at maximum heat in the fuel bed melting zone, and the oxygen entering the bed from the tuyeres must be entirely converted into carbon dioxide before coming into contact with the metal. If the bed is too low and the metal melts too close to the tuyeres, it will oxidize badly. If the melt takes place too high, it will be cold.

The two ways of determining if the melting is being accomplished correctly is by noting the time it takes for the first metal charge to melt, and by observing the color of the slag. If an apple green is observed upon fracturing the slag, things are going great and you can be assured that the melting is being accomplished in an area devoid of free oxygen. If the slag is dark brown, it contains some oxides of metal. If black, the metal is melting much too low.

What can be done to correct an oxidizing condition? Examine what's happening. The metal is oxidizing because there is free oxygen in the melting zone. This is caused by incoming oxygen not having time to be converted completely to carbon dioxide before reaching the melting metal. This can be due to too low a fuel bed, a too high blast pressure or both. To correct this condition, first lower the blast pressure, put in an extra 8-inch charge of coke, then add a normal metal charge, followed by 4 inches of coke when the 8-inch *split* reaches the bed. This can be determined by comparing the amount of metal tapped to the size of the metal charge. If the bed is too high, reduce the amount of coke between two charges of metal. When this coke reaches the bed, it will be lowered by this amount, or raise the blast pressure causing the melting zone to move up. When tapping the metal, watch for excessive sparks (oxidized metal) shooting from the stream issuing from the tap hole, indicating a low bed. The droplets of iron passing by the tuyere peep hole should be bright (hot) without giving off sparks. If they are dull

Fig. 5-23. Punching out chilled slag.

(cold) the bed is too high, the blast too low or both. If the droplets are bright, but shoot off sparks, the bed is too low, the blast is too high or both.

After you have done some actual cupola melting and trained your eye to recognize these symptoms, it will all become quite simple. If there is a foundry in your community that runs a cupola, try to get permission to look it over. Watch it operate and talk to the cupola tender. It could be a valuable outing.

When melting bronze, use a low blast pressure (2 ounces average, 3 ounces maximum) and a metal-to-coke ratio of 20 to 1. The tapping temperature should be approximately 2300°F, but no more. If the drops of bronze going past the tuyere peep holes are bright and clear, you are in good shape. If dull, they have become oxidized because of a low fuel bed, excessive blast or both. If they look cold but don't exhibit signs of oxidation, the bed is too high, the blast is too low or both.

When you have reached the end of your melting schedule, tap out all the liquid metal, shut down the blast, open a tuyere hole, pull the bottom door prop out and allow the burning coke and whatever else is left in the cupola to drop out. Quench the residue with a water. When the cupola is cold, you can tell just where the melting was being done by observing the groove burned completely around the refractory lining. The lining must be chipped clean of slag and patched before the next heat. Save the unburned, quenched coke from the *drop*, as well as any unmelted metal, for the next heat. Do not use drop coke in the next bed of coke. Use it between metal charges.

JAMMED TAPS

When a cupola is melting and cannot be tapped, if the blast is shut down, it stops melting at once and begins to freeze up. If the blast is continued, it continues to melt and if not tapped, the metal spills over into the safety tuyere or into the wind belt.

You can also lose your blast due to a mechanical or electrical breakdown and find that you cannot get the tap open and drain the stack due to a freeze up of the breast or a chunk of coke jammed in the way of the tap hole. To open a frozen or jammed tap, the simplest method is to burn your way through with an oxygen lance which consists of a length of black iron pipe attached to an oxygen bottle and regulator (Fig. 5-24).

A ¼ inch pipe 8 to 10 feet long is needed to operate the lance. The business end is heated with a welding torch until it just starts to

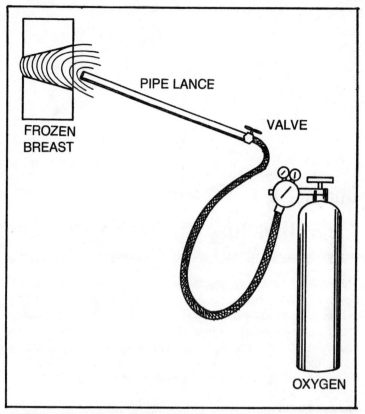

Fig. 5-24. To open a frozen or jammed tap, the simplest method is to burn your way through with an oxygen lance.

melt and the oxygen is turned on. This will support combustion at the end of the tube which is rapidly oxidizing with intense heat. The lance is used to burn a hole through the breast by melting anything in its way—refractory metal, etc.

It is common practice in some shops to provide two tap holes in the breast, one above the other. The top one is only opened in case of a freeze of the bottom tap.

Chapter 6
Metal Handling and Pouring Devices

There has been general appeal made by foundrymen, equipment designers and manufacturers for maximum safety in metal handling devices such as tongs, pouring shanks, etc. I am sure this is done with good intentions. However, there is still some metal handling equipment on the market that is not only cumbersome, but presents a hazard due to complicated construction and extremely poor balance that actually defeats its intended purpose.

CRUCIBLES

Crucibles are ceramic pots made of graphite and clay, silicon carbide or other refractories. They are used for melting and pouring metal, or as a ladle for pouring metal.

Melting and pouring crucibles are standardized by size, and numbered by the various manufacturers who also make special-purpose crucibles (Table 6-1). The crucibles most used fall into a range of sizes from a No. 10 to a number 400, representing 30 to 1200 pounds capacity for bronze and 10 to 400 pounds for aluminum. Generally a crucible will hold a weight three times its number of bronze or a weight equal to its number of aluminum; that is, a No. 20 crucible has a capacity of 60 pounds of bronze or 20 pounds of aluminum.

Crucible melting was practiced before the Christian era. Historical records mention crucible melting as far back as 5000 B.C.

The conventional pit or crucible furnace is still popular. The majority use a forced draft blower and burner, but a few natural-

Table 6-1. Crucible Sizes and Weights.

Number	HEIGHT OUTSIDE INCHES	TOP DIAMETER OUTSIDE, INCHES	BILGE DIAMETER OUTSIDE, INCHES	BOTTOM DIAMETER OUTSIDE, INCHES	BRIMFUL CAPACITY, CUBIC INCHES	APPROXIMATE WORKING CAPACITY, POUNDS RED BRASS
6	6½	5¼	5¼	3⅞	66.4	19.0
8	7⅛	5⅞	5⅞	4¼	91.3	26.3
10	8¼	6¼	6¾	5	125	35.6
16	9¼	7½	7½	5½	238	68
20	10½	7⅝	8-5/16	6⅛	285	82
30	11⅝	8-9/16	9-5/16	6⅞	437	125
40	12⅝	9⅜	10-3/16	7⅝	581	166
50	13-13/16	10¼	11⅜	8¼	664	190
60	14-9/16	10⅞	12-3/16	8⅞	775	221
70	15¼	11⅜	12-11/16	9¼	858	245
80	15-13/16	11-9/16	12-15/16	9-7/16	997	285
90	16⅜	12⅛	13-5/16	9¾	1052	301
100	16-13/16	12-7/16	13¾	10	1218	348
125	17½	12⅞	14⅛	10⅜	1383	387
150	18½	13¾	15⅛	11	1633	467
175	19⅜	14⅜	15¾	11½	2047	587
200	20¼	15⅝	16¾	12½	2210	634
225	20⅞	15-7/16	16¾	12½	2570	736
250	22½	16	17-3/16	12⅝	2680	769
275	22¼	16¾	18-3/16	13⅜	3040	872
300	22¾	16¾	18-3/16	13⅜	3300	943
400	24½	18½	19⅝	14½	4270	1219
430	25	20½	None	14½	5100	1457
430	25	20½	None	14½	5460	1561
500	22½	22⅜	None	14½	5460	1561
600	27⅜	23⅜	None	14½	6530	1869
700	32	23⅜	None	14½	8170	2338
800	36	23⅜	None		9420	2693

draft coke jobs are still around. The following guidelines are provided to insure the most efficient use of a crucible you might purchase, its longevity and your safety in handling it:

- When receiving a new crucible, inspect it carefully for damage in shipment and manufacturing defects. Tap the crucible lightly with a wooden club to check for internal cracks. If the crucbile is good, it will produce a clear ring—if not, a dull thud.
- Store crucibles in a warm, dry place with room for air circulation between them.
- Roll crucibles by tipping and rolling as you would a barrel, or carry them or use a hand truck.
- Use a base block made of the same material as the crucible. The base block should be at least as wide as the crucible.

Fig. 6-1. Come-out tongs should fit the bilge of the crucible properly, without being too close to the rest of the crucible.

- Place a piece of cardboard between the crucible and the base block to prevent the crucible from sticking to the block during melting. A stuck crucible has to be rocked to get it loose which is hard on the crucible, often tearing chunks out of its bottom.

- Do not wedge metal into the crucible. Leave room for metal expansion.

- Do not let the flame cover the crucible. The flame must be aimed from the side, at the line between the base block and the crucible bottom.

- Do not use excessive flux, or one that might be injurious to the crucible. Add the flux after the metal has been reduced to a semiliquid.

- Do not melt very small amounts of metal in a large crucible. This reduces crucible life.

- Do not allow a ridge of slag and metal oxides to build up in a crucible. It can cause gas bubbles in a casting known as *duck flights*: a trail of progressively smaller indentations caused by bubbles. Oxides will be dislodged from the ridge by molten metal and carried out with the pour. The oxides enter the mold cavity along with deoxidized metal, and start to give up their oxygen as they convert back to liquid metal, leaving a trail of bubbles which may not get out of the metal as it solidifies.

- Never clean crucibles with a chisel and hammer. Scrape gently while hot with a skimmer bar.

- Do not wedge blocks all around a crucible in a tilting furnace (in which the crucible is not removed). It must be allowed room to expand.

- Do not pull a crucible with tongs that do not fit properly, or grab the crucible above the bilge.

110

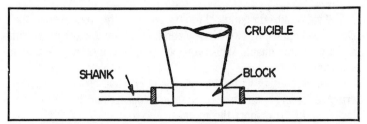

Fig. 6-2. Before a pouring shank can be attached to a crucible, the crucible must be resting on a block that is as wide as the crucible's bottom, and taller than the shank ring is wide.

- Do not use *come-out* tongs (Fig. 6-1) even when they fit properly at the bilge if they do not have sufficient clearance at the top of the crucible.

- Always use hand shanks and bull shanks that fit the crucible just below the bilge.

- Do not use, or attempt to make, homemade crucible tongs or pouring shanks. It is impossible to make this equipment properly without the proper jigs, experience and equipment necessary to do so. All it takes to cause a bad accident is to have an improper weld let go, or an improper fit to cause a dangerous spill. Besides, this equipment will cost you more in time and money to build than to buy.

- Never place a hot crucible in a draft. Return it to the furnace to cool slowly to avoid thermal shock.

- When attaching a pouring shank to a crucible, place the crucible on a base block as wide as the bottom of the crucible and extending above the shank ring (Fig. 6-2).

- Do not use wedges to make a shank accommodate a crucible smaller than it was intended for.

- Never use tongs with single grabs (Fig. 6-3).

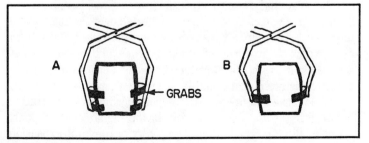

Fig. 6-3. Double-grab tongs (A) and single-grab tongs (B).

Two hundred pounds of molten metal is the most weight that should be handled by two men by hand; forty pounds, the most with a one-man hand shank. When you exceed this weight, use a crane or a jib boom.

When melting with a tilting-type of crucible furnace, a tapped crucible, or any open flame or electric furnace, pour the metal into a ladle used to transport the metal to the molds.

Ladles like these are made in various thicknesses of steel, depending on their size. Bowl-type ladles for a one-man hand shank are made of 14-gauge steel with a bead to catch the shank ring. The two most common sizes are type 14, and type 7. Ladles that fit two-man shanks range from 60 pounds capacity to 250 pounds capacity.

Ladles must be lined with a refractory and properly dried. Small ladles are usually lined with a castable or tamped-in commercial refractory such as A.P. Green, Red X or Green X ladle lining. A freshly lined ladle should be oven dried completely at a maximum of 300°F before putting it into service. Large ladles are lined with firebrick and faced with a tamped-in refractory.

A ladle must be brought to a red heat before tapping metal into it. This can be done by placing it on bricks above the furnace exhaust hole during the melting. Larger ladles, which would be cumbersome to heat in this manner, must be heated with a ladle heater.

The prices of crucibles have skyrocketed since 1970 therefore, it is a good idea to take good care of them to reduce costs and get the maximum life from them.

Store crucibles in dry places, especially ones made of graphite. Do not let them become damp by placing them on concrete floors or on the ground. Graphite crucibles should always be brought up to their operating temperature slowly. Also, never put a cold crucible in a hot furnace or turn the burner flame directly onto a cold crucible. Fire up the furnace for a few minutes. Then shut it off and place the crucible in so that it warms gradually from the residual heat of the furnace, or, the cold crucible can be inverted over the sight hole of the furnace if it is already operating. This procedure will warm and dry the crucible properly and gradually. None of these precautions are necessary with silicon carbide crucibles because they do not absorb moisture and are more resistant to thermal shock.

Cold chunks of metal should never be jammed into a crucible before firing it up. The expansion of the metal could crack the crucible. Place pieces loosely with plenty of room for expansion. You should never let a molten charge of metal solidify in a crucible.

Pour out almost all of the melted charge before it solidifies. The remelting of a solidified charge may cause the crucible to crack from the expansion of the metal.

Never use a crucible which shows any signs of cracking. It could break while lifting or pouring.

Lifting the crucible out of the furnace and pouring the molten metal is the most critical and dangerous operation of metal casting. The hobbyist must wear heavy leather or asbestos gloves, leather shoes and a face shield. He should also wear a heavy leather or asbestos leggings, preferably. Foundry suppliers offer a variety of safety equipment of this type, and it is the best investment that can be made by the home shop metalcaster.

A crucible should never be filled excessively. The level of molten metal should never exceed about 4/5 of the maximum capacity of the crucible. Special tongs should be used to lift and pour. No makeshift equipment should ever be used by any metalcaster.

TONGS

Crucible tongs are made in several designs and styles. They can be purchased from foundry suppliers, or you can make your own based on the designs shown in Fig. 6-4. The arc-shaped end which grips the sides of the crucible must fit its contour firmly and securely. Therefore, a different size tong will be required for each size of crucible. The L-shaped style shown in Fig. 6-4 seems to be safe, although this kind is more difficult to construct.

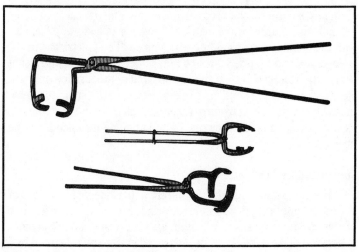

Fig. 6-4. Commercial tongs for lifting and handling a small crucible.

Color	Temperature°F.
Faint red	900
Blood red	1050
Dark cherry red	1150
Medium cherry red	1250
Cherry red	1450
Bright red	1550
Orange	1700
Yellow	1850
Light yellow	2100
White	2300

Table 6-2. Temperature Color Scale.

TEMPERATURE DETERMINATIONS

Many hobbyists will be tempted to judge the proper pouring temperatures of their metals by the appearance of the melt alone. This procedure can be used, but it is not suggested.

The *approximate* temperature inside the furnace or crucible can be estimated by looking through the sight hole in the lid and comparing the color with the scale in Table 6-2.

One problem in the use of such a table is the subjective interpretation of color. The amount of illumination may also be a factor. What looks like bright red in a home shop corner or on a dark day may appear as dull red in a brightly lit environment.

Optical Pyrometer

How does one measure the temperature of molten metals far above the range of thermometers? One instrument for such uses is the optical pyrometer, but it costs a couple hundred dollars so it is often beyond the reach of typical hobbyists. The optical pyrometer depends on comparing and matching the color of a glowing filament with the color of the object being measured. The electrical current is observed and adjusted through the filament until its image disappears when viewed against the background of the furnace or crucible interior. The filament and object are then at the same temperature. It can be read from a calibrated dial on the instrument. Readings can be taken very rapidly and at some distance from the furnace.

Thermocouple Pyrometer

For measuring high temperatures, a thermocouple pyrometer is wed. Its operation depends on the principle that when the junction of two dissimilar metals is heated, an electrical voltage becomes generated. This voltage can be measured in an external circuit by a device such as a meter or potentiometer. The voltages (or cur-

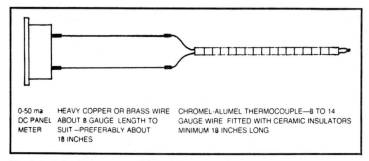

Fig. 6-5. Schematic of a simple thermocouple pyrometer.

rents) generated are quite small but they are measurable with simple equipment.

Connect a thermocouple (the two pieces of dissimilar metal in the form of a pair of wires) to a simple direct current meter. The current and the meter reading depend not only on the temperature of the junction, but also on the resistance of the meter and wires, therefore, a simple instrument of this kind does not measure temperature directly. It is calibrated by comparing the reading at known temperatures.

Buy a 0-50 milliamp direct current meter from an electronic or surplus store. It should have a scale length of at least 2 inches for accuracy and be readable to 1 milliamp or better.

Also obtain a chromel-alumel thermocouple made of 8 to 14 gauge wire. It should be equipped with ceramic insulators and with a minimum length of at least 18 inches. Then connect the two leads of the thermocouple through the intermediate heavy copper or brass wire (Fig. 6-5). The connection from the thermocouple wire to the lead wires should be brazed or silver soldered, although good mechanical connectors should work. Tight, secure electrical connections are essential for stable readings of the instrument. The lengths of the intermediate lead wires can be made to suit, but about 18 to 24 inches should suffice for readings in a small furnace.

The thermocouple pyrometer is now ready for calibration. Should the meter needle deflect downscale when the tip of the thermocouple is heated, reverse the leads to the meter. The meter should read zero when the thermocouple is at room temperature. If not, adjust it to zero with the small screw below the face of the meter.

Chapter 7
Metallurgical
Properties of Cast Metals

Pure metals are rarely used for making castings. The properties of pure metals can be improved with respect to fluidity, melting point, strength and hardness by the addition of alloying elements. Metallurgy has resulted in the development of certain standard casting alloys which have ideal properties for given applications, and the hobby metalcaster will find that the use of these materials will yield favorable results.

A particular metal or alloy will be chosen for a given casting dictated by a number of factors such as required strength, surface finish desired, cost, ductility and ease of machining and casting. Cast iron does have certain deficiencies, but its low cost offsets the disadvantages.

POPULAR ALLOYS

Many aluminum alloys are used for metal casting. The most versatile are the aluminum-silicon alloys which combine good castability with excellent corrosion resistance and high strength.

Aluminum

Aluminum alloys cast very easily and the beginner is encouraged to experiment with aluminum casting first before graduating to the more difficult-to-handle brasses and bronzes. Basically the techniques are the same in all cases, but aluminum alloys are easier to handle and melt.

Scrap aluminum works well if you carefully pick out only scrap castings such as gear cases, cylinder heads, machine parts, etc. Avoid pistons and connecting rods because they are alloyed with other elements which may make the material less suitable. Avoid wrought or structural aluminum scrap completely since it is not alloyed for casting purposes and will give poor results.

Segregating aluminum and magnesium alloys is not always done too well by scrap dealers. Do not mix magnesium with aluminum scrap because the accidental melting of magnesium may lead to a fire. Once ignited, magnesium cannot be extinguished. If it should ignite, cover it with sand to slow the burning and let it go. Never apply water to burning magnesium—only dry sand. Sometimes magnesium alloys can be identified by sight. It is grayer in color and is less dense than aluminum. If there is any doubt, apply a drop of 1 percent silver nitrate solution to the filed surface of the metal. A black stain will be produced immediately if the metal is magnesium. Aluminum and its alloys will remain unchanged.

Silicon Bronze

Silicon bronze casting alloys have grown in popularity. They have excellent casting properties with high strength approaching that of low-carbon steel and good corrosion resistance. Silicon bronze alloys contain about 95 percent copper and 4 to 5 percent silicon. They also contain minor amounts of manganese and zinc.

Silicon bronze has the best combination of properties which appeal to the amateur metalcaster of all readily available copper-based alloys. It is inexpensive and can be repeatedly remelted without changing composition. Silicon bronze is available under the trade names Herculoy and Everdur.

Red Brass

Red brass can be considered as either a brass or a bronze. The composition varies, but the typical red brass contains about 85 percent copper and 5 percent each of tin, lead and zinc. Red brasses are among the most widely used of all copper-based casting alloys. They have numerous uses in the production of pipe, valves, fittings, pump housings and plumbing fixtures. They also offer an excellent combination of resistance to corrosion, high strength and casting properties.

Bronze

Bronze is not an alloy of definite composition. It is an alloy of copper and tin. It may also contain small amounts of other elements.

Other bronzes must also be recognized which contain relatively little tin, such as manganese bronze, as well as other alloys which contain no tin at all, such as aluminum bronze (copper-aluminum) and silicon bronze (copper-silicon). Some alloys are called bronzes but in reality are brasses, such as architectural bronze (copper-zinc-lead) and commercial bronze (90 percent copper, 10 percent zinc).

The term *bronze* used without qualification usually means tin-bronze. When tin is added to copper it increases its hardness and strength. Zinc is sometimes added in small amounts to improve the casting properties. Lead improves the machining qualities. Tin bronzes contain about 87 to 90 percent copper, 6 to 10 percent tin and 2 to 4 percent zinc. If lead is added to improve machining, it will usually be present by about 1 percent. The alloy will then be referred to as a leaded tin bronze. Tin bronzes have excellent casting properties as well as good mechanical strength and ductility. Tin bronzes do tend to be a little more expensive than yellow brasses, however, which contains no tin.

Bronwite

Bronwite is a patented alloy manufactured by ASARCO. Originally it was developed to replace nickel silver alloys and to avoid the high cost and availability problems involved with nickel. Bronwite contains 59 percent copper, 20 percent zinc, 20 percent manganese and 1 percent aluminum. With respect to color, corrosion resistance, strength, ductility and hardness it is the equivalent of nickel silver, but it easier to cast because of the low melting temperature of 1550°F. Bronwite can be polished to a very high luster. Parts originally made of nickel do not need be plated if they are made of Bronwite. Also, its resistance to tarnishing is just as good as nickel.

Zinc

The most common alloys of zinc are the Zamak alloys which contain about 95 percent zinc, with 4 to 5 percent aluminum and occasionally slight additions of copper and magnesium. Zinc die casting alloys have good strength and casting properties, but they do lack corrosion resistance and ductility. The only advantage is the low melting temperature which reduces the high temperature capabilities required for the melting furnace.

UNDESIRABLE ALLOYS

Many alloys used during metal casting should be avoided by the beginning metalcaster in the home hobby shop.

Manganese Bronze

A favorite casting alloy for parts requiring high strength and resistance to corrosion is manganese bronze. Manganese bronze compositions vary quite a bit. They depend upon the strength characteristics desired, but typical compositions are around 60 percent copper and 25 to 36 percent zinc, together with small amounts of iron, aluminum and manganese.

Because manganese bronze is not an easy material to cast the amateur should avoid it.

Aluminum Bronze

Aluminum bronze has enormous strength, far exceeding that of mild steel. In addition, it can be heat treated to improve its strength. It has a narrow solidification range high shrinkage upon solidification. Therefore, aluminum bronze is an extremely difficult material to cast.

Brass

Brass is readily available in scrap form which would appear to make it an attractive choice for the amateur. However, brass is not easy material to cast and the hobbyist is likely to experience some difficulties with it, especially at first.

Brass does not denote a specific composition of material. It is a class of alloy. Brass is an alloy of copper and zinc containing more than 50 percent copper. It may also contain minor amounts of other alloying materials. Yellow brass ordinarily contains 65 percent copper and 35 percent zinc. It has a pleasing yellow color. It also polishes well.

The term *brass* used without qualification usually means yellow brass. It contains copper and zinc. Other compositions are also used for specific applications in industry. Scrap brass contains a variety of different compositions mixed indiscriminately. The metalcaster using scrap brass will find that casting properties vary from melt to melt.

When brass is brought to its melting temperature, zinc distills out. It also burns above the molten alloy. This problem can be alleviated by providing a protective cover of flux over the melt.

SCRAP AND VIRGIN ALLOYS

Scrap brass and bronze are readily available at scrap metal dealers. They are often at prices about half that of commercial alloy prices. A beginner may be tempted to buy scrap metals for his first castings, but this practice is not to be encouraged. Copper alloys are difficult to identify from appearance alone, so many alloys will end up in the same bin at the scrap dealers. You could end up with a mixture of yellow brass, aluminum bronze and manganese bronze. Needless to say, you would find uncertain results on the quality of the finished product. Select your material from a reputable dealer and reject any scrap with an uncertain color.

Casting alloys should really be purchased in ingot form. More uniform results and fewer rejects will make up for the higher raw materials costs. Brass and bronze are usually sold in 25-pound ingots, but they can be cut up into more convenient 4 or 5 pound chunks with a power hacksaw.

Chapter 8
Cores and Their Application

A core is a preformed baked sand or green sand aggregate inserted in a mold to shape the interior part of a casting which cannot be shaped by the pattern.

A core box is a wood or metal structure, the cavity of which has the shape of the desired core which is made therein.

A core box, like a pattern is made by the pattern maker. Cores run from extremely simple to extremely complicated. A core could be a simple round cylinder form needed to core a hole through a hub of a wheel or bushing or it could be a very complicated core used to core out the water cooling channels in a cast iron engine block along with the inside of the cylinders.

Dry sand cores are for the most part made of sharp, clay-free, dry silica sand mixed with a binder and baked until cured. The binder cements the sand together. When the mold is poured the core holds together long enough for the metal to solidify, then the binder is finely cooked from the heat of the casting until its bonding power is lost or burned out. If the core mix is correct for the job, it can be readily removed from the castings interior by simply pouring it out as burnt core sand.

This characteristic of a core mix is called its *collapsibility*. The size and pouring temperature of a casting determines how well and how long the core will stay together. A core for a light aluminum casting must collapse much more readily than a core used in steel because of the different time, weight and heats involved. Core sand, like molding sand, must have the proper permeability for the job intended. The gases generated within the core during pouring must be vented to the outside of the mold preventing gas porosity and a defect known as a core blow. Also, a core must have sufficient hot strength to be handled and used properly. The hot strength refers to its strength while being heated by the casting operation. Because of the shape and size of some ores they must be further strengthened with rods and wires.

A core for a cast bronze water faucet must have a wire in the section which passes through the seat of the valve. Because of the small diameter at this point, core sand alone would not have sufficient dry strength to even get the core set in the mold much less pour the castings (Fig. 8-1).

A long span core for a length of cast iron pipe would require rodding to prevent the core from sagging or bending upward when the mold is poured because of the liquid metal exerting a strong pressure during pouring (Fig. 8-2).

BINDERS

There are many types of binders to mix with core sand. A binder should be selected on the basis of the characteristics that are most suitable for your particular use.

Some binders require no baking and become firm at room temperature. Examples are rubber cement, Portland cement and sodium silicate or water glass. In large foundry operations and in some school foundries, sodium silicate is a popular binder as it can be hardened almost instantly by blowing carbon dioxide gas through the mixture.

Oil binders require heating or baking before they develop sufficient strength to withstand the molten metal. Linseed oil is considered one of the strongest binding oils. Vegetable, mineral and fish oils are also used.

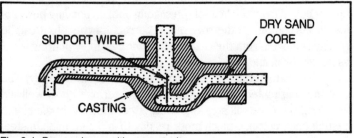

Fig. 8-1. Dry sand core with support wire.

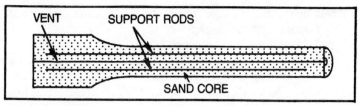

Fig. 8-2. Rodded core.

Sulfite binders also require heating. The most popular of the sulfite binders is a product of the wood pulp industry. It is sold in liquid form under the trade name of Glutrin and in the dry form under the trade name of Goulac.

Furan is the trade name of a chemical binder which is hardened by the use of a catalyst. There are many liquid binders made from starches, cereals and sugars. They are available under a countless number of trade names.

A good binder will have the following properties:

● Strength
● Collapse rapidly when metal starts to shrink.
● Will not distort core during baking.
● Maintain strength during storage time.
● Absorb a minimum of moisture when in the mold or in storage.
● Withstand normal handling.
● Disperse properly and evenly throughout the sand mix.
● Should produce a mixture that can be easily formed.

CORE MIXES

The following is a list of core mixes that I have used for a number of years with good success.

For brass and aluminum

Fine river sand	10 parts
Wheat flour	¼ parts
Air float sea coal	3 parts
Rosin	1 part

For small castings

New molding sand	15 parts
Sharp sand	5 parts
Linseed oil	1 part

For aluminum, heavy castings

Sharp sand	30 parts
New molding sand	10 parts

Wheat flour..2 parts
    Temper with molasses water.
For aluminum, medium castings
    Sharp sand ....10 parts
    New molding sand ....5 parts
    Wheat flour ....1 part
    Temper with molasses water.
Good mix for long skinny cores
    Sharp sand ....8 quarts
    Wheat flour...1 quart
    Core oil ....⅛ pint
Quick collapse cores for aluminum
    Sharp sand ....45 parts
    Molding sand ....45 parts
    Powdered rosin..2 parts
    Wheat flour ....1 part
Mix for small cores
    Sharp sand...25 quarts
    Molding sand..15 quarts
    Linseed oil..1 quart
Brass pump core mix
    Silica..32 parts
    Fire clay...1 part
    Silica flour...4 parts
    Add rosin to desired dry strength.
Small diameter bushing cores
    Sharp sand ....80 parts
    Corn flour...2 parts
    Oil ....1 part

If you are now confused about core mixes, don't let it worry you. Foundry A produces the same type and weight of castings as Foundry B. Foundry A has 100 or more different core mixes. They use almost a different core mix for each job. Foundry B doing the same kind of work uses three different mixes, one for light work, one for medium work and one for heavy work.

Both produce good work. However, I am sure Foundry B's cost and problems are much less than A. This condition exists throughout the industry. For general all round core work in the small or hobby shop, a simple linseed oil bonded mix will suffice for 90 percent of the work.

The simplest mix is by volume, 40 parts fine sharp washed and dried silica sand and one part of linseed oil.

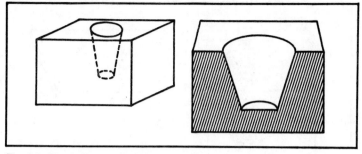

Fig. 8-3. Simple dump core boxes.

It is common practice to add a small percentage of water and kerosene. This makes it mix better and strip easier from the core box.

If the moisture is too low in the mix the core will bake out too soft and if the moisture is too high the core will bake out too hard and stick in the box when making the core. Most mixes work best when just enough water is added to make the mix feel damp but not wet. To mix your core mix start out with dry sand, nearly pure silica sand free from clay. Mix the dry ingredients, add the oil and water and finish mixing.

Core sand mixes can be mixed in a muller or paddle type mixer and in small amounts on the bench by hand.

The core is made by ramming the sand into the core box and placing the core on a core plate to bake (Fig. 8-3).

The box cavity is dusted with parting powder usually made of powdered walnut shells, purchased as a core box dry parting. The box is rammed full of sand using the handle of a raw hide mallet to ram. The excess is struck off level with the side of a core maker's trowel. The core is then vented with a vent wire, a core plate placed on top and the plate and box rolled over on the bench. The box is rapped on one side and then on the front and back to loosen the core.

The box is then lifted (drawn) off leaving the core on the plate. The plate is placed in the oven to be baked or dried. When finished, it is removed from the oven and allowed to cool. It is then ready to use (Fig. 8-4).

Stand-Up Cores

Stand-up cores (cores that can be stood on end to dry) are made in a split core box. The box is pinned together with pattern bushings. The box is held together with a C-clamp and it is rammed up using a dowel of suitable diameter. When rammed, it is then

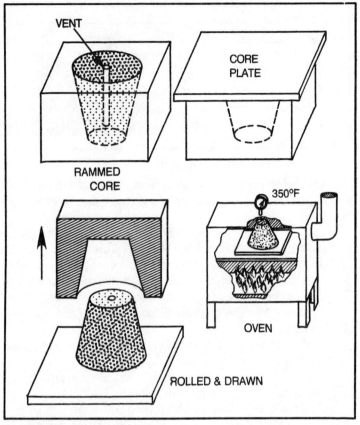

Fig. 8-4. Use of dump core box.

vented with a vent wire. The clamp is removed and the box placed on a core plate. The half of the box away from the core maker is removed and the core is slid to the far side of the plate. The remaining half of the box is removed, leaving the core standing free. This is continued until the core plate is full, or until you have sufficient cores. The plate is then ready for the core oven (Fig. 8-5).

Stand-up cores are often made in gang core boxes (Fig. 8-6).

Stock or round cores, when too tall to stand on end, are made in a split box and rolled on to a core plate for drying (Fig. 8-7).

Pasted Cores

Cores can be made in halves and after they are dried, glued together to make the complete core. If the core is symmetrical, a half box is all that is needed (Fig. 8-8).

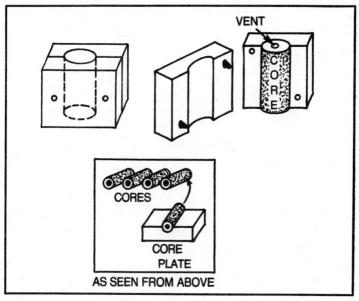

Fig. 8-5. Stand-up core box.

After the two halves of the core are dried, a vent is scratched along the center line of one half and the sections are glued together with core paste. The seam is then mudded with a material known as *core daubing*. Both core paste and daubing can be purchased or made. For homemade core paste use enough wheat flour dissolved in cold water to produce a creamy consistency.

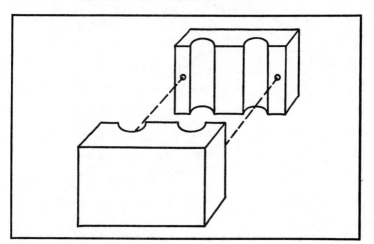

Fig. 8-6. Gang core box.

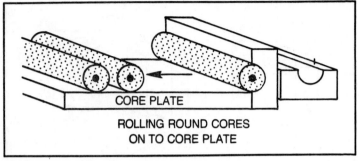

CORE PLATE

ROLLING ROUND CORES
ON TO CORE PLATE

Fig. 8-7. Use of core plate.

For a good core daubing mix fine graphite with molasses water. (One part molasses to 10 parts water.) Enough graphite is added to the molasses water to make a stiff mud.

Another good daubing mix is graphite and linseed oil mixed to a stiff mud.

After a core has been pasted and daubed, it is a good idea to return the core to the core oven for a short period to dry the paste and daubing.

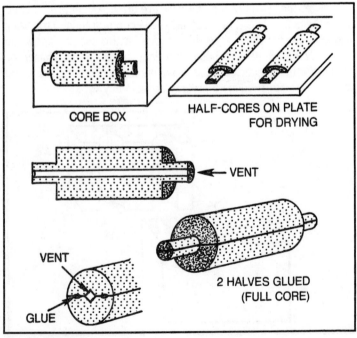

CORE BOX

HALF-CORES ON PLATE
FOR DRYING

VENT

VENT

GLUE

2 HALVES GLUED
(FULL CORE)

Fig. 8-8. Pasted cores.

Three Part Core Box

This box consists of a top, front and back section. The box is assembled and placed top down and clamped. You ram it up, put on a core plate, roll it over and rap it. Remove the top (which forms the negative section on top of the core), then remove the main box from the core as you would any simple split box (Fig. 8-9).

Loose Piece Box

A loose piece box is a box which contains a loose piece (or drawback) to form a particular shape to the core. The loose piece is placed in position in the box, the core rammed, rolled over on a plate, rapped and lifted off. The loose piece remains with the core on the plate and is drawn back to remove it (Fig. 8-10).

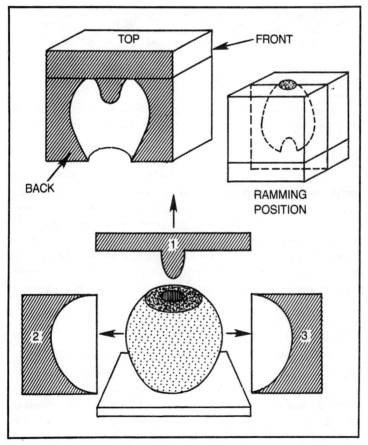

Fig. 8-9. Three part core box.

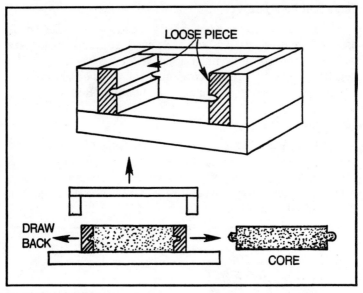

Fig. 8-10. Loose piece core box.

The core box may have one loose piece or many. Sometimes a loose piece may have a loose piece of its own.

Swept Core

A core can be produced by a suitable shaped sweep (or stickel) and guide rails. The guide rails are clamped to a core plate. Core sand is heaped in between the rails and swept to shape by the sweep form. When finished the guide frame is removed and the core dried. The swept core might have several different diameters and may be curved. The shape dictates the frame construction and the number and shape of the strikes. Flat cover or slab cores are more often than not swept up on a core plate with a simple frame and flat strike off bar (Fig. 8-11).

CORE DRIERS

A core drier is a support used to keep cores in shape during the baking or drying. Most core driers are made of cast aluminum. The core drier is similar to a half core box but fits the core more loosely than the box and only sufficiently to prevent the core from collapsing or sagging before and during drying. The core is rammed in the usual manner. The side of the core box that presents the face of the core that will rest in the drier is removed. The drier replaces this

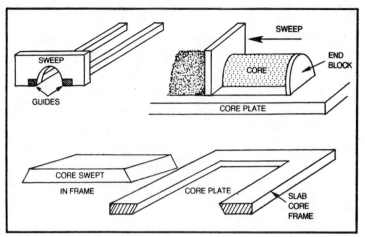

Fig. 8-11. Swept core.

half box. The box and drier are rolled over and the box removed, leaving the core resting in the drier ready to be dried (Fig. 8-12).

In the case of a green sand drier, you simply make a bed of riddled molding sand to support the core during drying. The core box is rammed, one half of the box removed and replaced by a tapered wooden frame. Dry parting is shaken over the exposed half of the green, the unbaked core is filled to the top with lightly riddled, tempered fine molding sand and struck off level with the top of the frame. A core plate is then placed on the frame and the whole

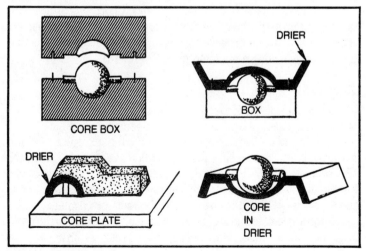

Fig. 8-12. Core driers.

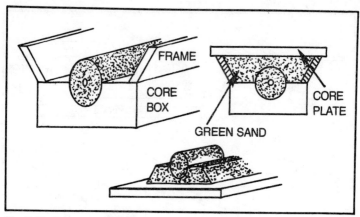

Fig. 8-13. Green sand drier.

assembly rolled over. The remaining half of the core box is re-moved, then the wooden frame. This leaves the core supported in a bed of molding sand for drying. When dry, the green sand is simply brushed off of the baked core with a molder's brush (Fig. 8-13).

BALANCE CORE

A balance core is when the core is supported on one end only and the other unsupported end extends a good way into the mold cavity. In order to prevent the core from being lifted up by the metal coming into the mold we must extend both the length and diameter of the core section held by the single print (Fig. 8-14).

To prevent having to make a balanced core which not only takes up room in the flask, but increases core cost, a better solution is to use one core to produce two castings. It is a core to produce two castings having a common print (Fig. 8-15).

CHAPLETS

Chaplets consist of metallic supports or spacers used in a mold to maintain cores, which are not self supporting, in their correct position during the casting process. They are not required when a pattern has a core print or prints which will serve the same purpose. Chaplets are purchased in a wide variety of sizes and shapes to fit just about any need or condition. The three most common types are the stem chaplet, motor chaplet and the radiator chaplet.

Motor chaplets and stem chaplets are placed or set in the mold after the mold is made. Radiator chaplets are rammed up in the mold during the molding operation.

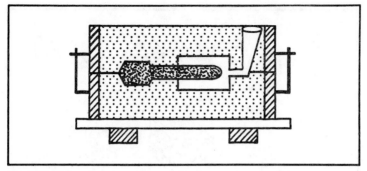

Fig. 8-14. Balance core.

On many gas stove burners or hot water heater burners and many types of gas furnace burners, you will note that there are a series of projections or little bumps on the casting face both on the cope and drag. They are more noticeable on the bottom face. These bumps or knots are made by radiator chaplets used to hold the core centrally located when these burners are cast. The pattern is drilled wherever a chaplet is needed. These drilled holes are slightly larger in diameter than the stem diameter. The chaplet has a knot back from the business end of its stem that allows the chaplet to only stick into the hole in the pattern as far as this knot. The distance from the knot to the end of the stem represents the metal thickness of the casting. The end of the chaplet that is held in the molding sand has a plate shaped head to give it a good purchase in the rammed molding sand which prevents it from moving or being pushed back by the force of the core as it tries to float when the mold is being poured.

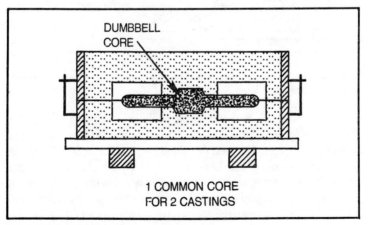

Fig. 8-15. Dumbbell core.

133

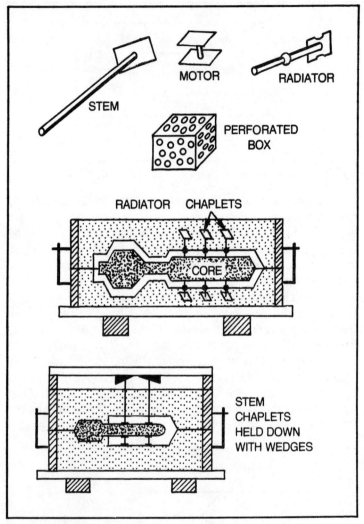

Fig. 8-16. Chaplets.

The molder sets the stem chaplets into the holes in the drag half of the pattern. He rams the drag, rolls the flask over and sets the cope chaplets. He then rams and finishes the cope. When the pattern is removed you have the stems of the chaplets sticking through the molding sand the exact distance required. The core is set on the drag chaplets and the chaplets in the cope come down and clamp the core between the cope and drag chaplets when the mold is closed.

When the casting is poured, the projecting chaplets hold the core in place until the metal starts to solidify, then these projections fuse or weld themselves into the metal that surrounds them.

A leaking hollow casting sometimes results when using radiator chaplets. This problem arises when the casting is not poured hot enough to properly fuse the chaplet, or a rusty or dirty chaplet is used by the molder. After the casting is shaken out, the stem and head that was in the sand is broken off. The chaplet is provided with a break off notch (Fig. 8-16).

Chaplets are seldom, if ever, used in nonferrous casting because the pouring temperature is not high enough to melt and fuse the chaplet.

A core which is set in place during the ramming of a mold to cover and complete a cavity partly formed by the withdrawal of a loose piece on the pattern is called a cover core.

RAM-UP CORE

A ram-up core is a core that is set against the pattern or in a locator (slot, etc.) in the pattern. The mold is rammed and when the pattern is drawn, the core remains in the mold (Fig. 8-17).

As you can see by now there is an endless variety of types, kinds and uses of cores. New uses and kinds are continuously coming up as new problems present themselves.

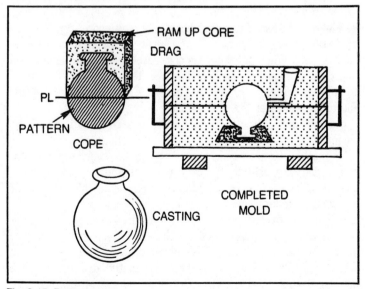

Fig. 8-17. Ram-up core.

CORE WASHES

Cores are sometimes coated with a refractory wash to increase the cores' refractiveness and to produce a smoother metal surface in the cored cavity of the casting. These materials are called *washes*. They can be purchased in a variety of types and refractive strengths. A common homemade wash is graphite and molasses water mixed to a nice paint consistency.

CORE PLATES

Core plates as you have no doubt guessed by now are flat plates used to bake the cores on in the core oven. They can be made of cast aluminum which has been normalized and machined to a true flat surface. Composition plates can be purchased which are called Transite, a composition asbestos mill board manufactured by Johns Manville Co. They can be had in a range of sizes and thicknesses from any foundry supply house.

Vent waxes are tapers of various diameters made of a wax and rosin combination. The vent wax is rammed up in the proper position in a core, and when the core is baked, the wax melts leaving a vent channel. This wax is sold in spools by supply houses.

CORE OVENS

The cores are baked in order to set the binder. The usual temperature range for oil bonded cores is from 300 to 450 degrees F. The time required varies with the bulk of the core. A large core might take several days to bake or a small core might bake out in an hour or less. When an oil core is completely baked the outside is a rich dark brown, not black or burned. The core must be cured completely through with no soft centers. Only experience and trial and error will teach you how to bake cores.

It is common practice to hollow out large cores to decrease the baking time and assure an even bake throughout.

Another factor which relates to the time and temperature required to properly dry a core is the type and amount of binder used. Oil binders require hotter and quicker baking than rosin, flour and goulac binders.

The core oven, which is usually a gas fired oven with temperature controls, is equipped with shelves on which to set the core plates and cores for baking. Some have drawers like a file cabinet which pull out to load and unload. When the drawer is pulled completely out, its back closes off the opening in the oven and prevents heat loss. This type of oven is also made with semicircular shelves which swing out.

The core oven can consist of a square or rectangular brick oven with doors. The bottom of the oven is floor level. The cores are placed on racks which, when full, are rolled into the oven. The oven is closed and the cores are baked. It is common practice to load the ovens one day with the cores required for the next day and to dry them over night. The lead time required for cores to the molder will vary from hobby shop to shop.

You can bake cores in your kitchen oven or there are many different kinds of mechanized machinery which can be purchased to make cores, completely automated to semi-automated, blowers, extruders, shell machines, etc.

SHELL COREMAKING

Shell cores are, like shell molding, becoming more and more popular for certain types of work. Shell cores are coming to the front very rapidly. One of the most important things to consider in designing with a shell core is to use it when it offers a definite advantage over an existing dry or green sand core.

Carefully consider the casting, its design, etc., to determine whether a decided advantage can be realized by use of a shell core. Look for an overall reduction in cost, less machining, elimination of an operation, etc. Don't use a shell core to core out something like the hole in a frying pan handle or some such foolishness as that.

Sometimes a combination of shell and dry sand cores in the same job is the best practice. Use the methods and types of materials you have to the best advantage. Avoid foolish or ridiculous applications. Speaking of the latter, I have seen some errors, but I also have seen some very ingenious applications of shell cores which resulted in concrete advantages and dollar savings.

So far as I know, there is no difference in the sand, binder or method employed to make a shell core for use in nodular or any other type of metal, whether ferrous or nonferrous. I have seen shell cores used to make a gray iron casting, then used to produce the same job in brass with no change in equipment or cores. It is my understanding, however, that the most difficult metals to cast in conjunction with shell cores are certain classes of steel and the copper-base alloys with more than 6 percent lead.

A shell core sometimes has a tendency to be thin at the vertical surfaces which are at right angles to the body; that is, the sides of the core that forms the sides of your rings. However, with a ½-inch

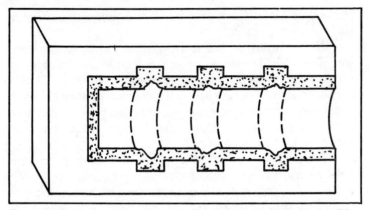

Fig. 8-18. Ring projections on the core fill up with sand and become solid.

wall in the casting, the ring projections on the core undoubtedly will fill up completely with sand and be solid (Fig 8-18). Using a thin, hollow shell core over a long unsupported span completely surrounded by metal is like trying to duck a balloon under water. If you have no cores rods or other stiffening or supporting members, the lifting force on the core is somewhat greater because of its lightness and its hollowness. I have seen shell cores reinforced internally with molding sand or just plain dry, sharp sand.

The print on the open dump end of some cores should be long enough to insure a good seal to prevent metal from bleeding or running into the core. Most hobby shops using shell cores cut a groove around the core prints, a little way back from the casting (Fig. 8-19). This procedure will give you a little green sand crush to form a tight seal around the core print. Dough roll, core paste or one of the prepared seals with a little corn flour sprinkled on makes a good seal.

The percentage of resin needed in the mix seems to vary from sand to sand, increasing with sand fineness. The usual amount is about 5 percent by weight, but there is a critical amount which must be worked out. As the amount of resin is increased, the resulting shell or core becomes smoother and harder. It is with this condition that you might run into penetration, cracks, etc. A high percentage of resin makes a hard and brittle core. It seems that above 8 percent resin, cores are too hard and brittle. Below 4 percent, they usually are too weak. I believe that the proper ratio of sand to binder will produce the desired results.

The sand and resin may be mixed dry in a suitable muller or a Y-type blender and used in that condition, or the sand may be resin coated. Sand may be coated with resin by either the hot or the cold process. Since the latter can be carried out easily in the small shop, I will touch on it briefly.

Sand and the proper amount of resin—about 2.5 pounds per 100 with round grain and 4 pounds with subangular sand—are mixed dry in a muller for 1 to 2 minutes. Then an alcohol-water solvent is added slowly. The solvent is three parts denatured ethyl or iso-propyl alcohol to one part water by weight. The amount varies with resin. It is 0.72-pound solvent for the 2.5-pounds resin mix and 1 pound solvent for the 4-pounds resin mix.

Mulling is continued for 10 to 20 minutes until the solvent has evaporated and the sand is dry. That condition is indicated by appearance and can be checked by picking up a handful of the sand. It should show no strength. Mulling is stopped as soon as the sand mix is dry. If you try this, don't forget that the muller should be hooded and hitched to an exhaust system.

This mixture is designed to coat every grain with a coating or shell of resin, preventing the possibility of segregation which some-time occurs in sand resin mixes. Any resin which is capable of heat polymerization can be used, such as urea formaldehyde or phenol formaldehyde. Use a top-quality, clay-free, washed and graded subangular silica. It should be below 2 percent clay AFS.

FALSE CORING OF BAS-RELIEFS

The field of art casting is for the most part a speciality and is full of tricks requiring lots of experience.

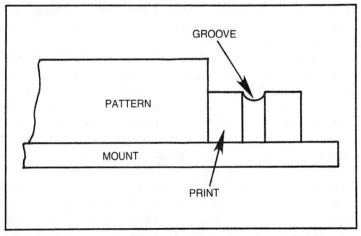

Fig. 8-19. Cut a groove around the core prints when using shell cores.

The system most commonly used for making a casting with undercuts, such as bas-reliefs, is called *false coring* (Fig. 8-20). Each recess or undercut is filled with sand in such a manner that it leaves a print on the main face of the mold where it can be attached. These portions have to be made so that the main face of the mold can be lifted from the face of the pattern without breaking or tearing up. At the same time, each false core must be made so that it can be removed from the pattern without damage to it and then be placed in its proper print on the mold face.

French sand is used extensively for this type of work because it can be rammed hard and can be dried hard without the fear of blowing, etc. French sand comes from a little town near Paris. *Yankee sand* is what often is called synthetic French. It is used widely for fine work. In fact, I have seen Yankee sand do just as well as French sand. With either sand, if the whole mold is put up, it should be rammed carefully and hard, washed with plumbago and oven-dried. If Yankee or French is used as a facing, skin drying alone will suffice.

French sand has a grain fineness of 140 to 175 with the clay removed, carries about 16 percent clay substance, has a permeability of 18 at 7 percent moisture and averages a green compression strength of about 10. With Yankee sand, grain fineness is about 184, clay content about 19.5 percent and permeability approximately 17. Green compression strength runs to an average of 16.

Should the bas-reliefs be flatbacked, as most of them are, the cope is made up flat on a smooth, level board. It is struck off and rolled over on a bed. The pattern is then placed on the cope in the correct position for ramming the drag with the rococo face up. The next move is to study the arrangement carefully and decide about the type of false cores necessary.

Each is made by carefully ramming into the undercuts, bearing in mind that they later must be removed and reassembled on the drag face of the mold in their correct position. Use wires and metal lifters where necessary. Often because one undercut is above another, one false core requires one or more false cores itself.

Make sure that each false core has a sufficient print to locate it correctly with respect to the finished drag. Each false core must have correct draft to allow the drag to be rammed up and removed in good shape. It should be remembered that if your piece has rococo and backdrafts on each side, the procedure is the same with the exception that you have to use an odd-size double roll, treating each side singly.

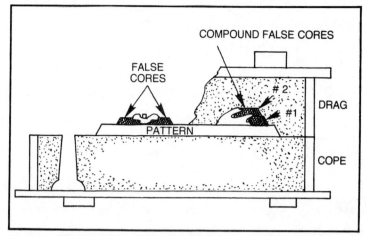

Fig. 8-20. Bas relief with a backdraft requires the use of false cores.

After all the false cores have been made, they are marked in such a manner that they will leave an identifying mark on the drag for relocation. The face is dusted with parting, and the drag flask is placed, carefully rammed up and struck off. A bottom board is rubbed in place. Drag and bottom boards are lifted off and rolled over on rails. The false cores then are finished from the undercuts and glued or nailed into their correct positions on the drag face. The rest is as usual—skin dry, oven dry or use whatever other system you prefer.

Some operators make up their false cores and bake them with infrared lamps just where they are, on the face of the pattern, before they ram the drag. This procedure can be used with French, Yankee or oil-sand cores provided that the pattern is made of a material which will take the heat. A wooden soldier here and there, cut so as to give both a print and a handle by which to handle the false core, sometimes can be used to an advantage.

I have seen some old French patterns carved in mahogany which were loaded with rococo and gingerbread. In turn they loaded with undercuts and backdrafts, but were only optical illusions. To create the illusion of folds and undercuts without actually having any is my idea of being really clever.

I also have seen the old cup, saucer and spoon made with false cores in a two-part flask. Try that some time.

Chapter 9
Patterns and Related Equipment

A pattern is a shaped form of wood or metal around which sand is packed in the mold. When the pattern is removed the resulting cavity is the exact shape of the object to be cast.

The pattern must be designed to be easily removed without damage to the mold. It must be accurately dimensioned and durable enough for the use intended.

Each different item we wish to cast presents unique problems and requirements. Especially in large foundries there is a close relationship between the pattern maker and the molder. Each is aware of the capabilities and limitations of his own field.

Throughout the industry, pattern making is a field and an art of its own. The pattern maker is not a molder nor the molder a pattern maker. This is not to imply that the pattern maker cannot make a simple mold or the molder make a simple pattern but each may soon reach a point in the other's field beyond his own skill and experience.

In the hobby or one-man shop, however, pattern and mold making are so closely interrelated that they become almost one continuous operation. This chapter will acquaint you with some of the various types of patterns and their requirements.

DRAFT

In order to illustrate some of the important pattern characteristics, we will use as an example a simple disc pattern. The object we want to cast is 12 inches in diameter and 1 inch thick. The edge of

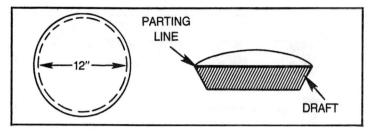

Fig. 9-1. Disc pattern.

the disc is tapered from the top face to the bottom face (Fig. 9-1). This taper is known as the *pattern draft*. This draft is necessary in order that the pattern can be removed easily from the mold causing no damage to the sand. Pattern draft is defined as the taper on vertical elements in a pattern which allows easy withdrawal of the pattern from the mold. The amount of draft required will vary with the depth of the pattern. The general rule is ⅛-inch taper to the foot which comes out to about 1 degree. On shallow patterns such as our disc 1/16-inch taper, 0.5 degree is sufficient.

SHRINKAGE

Now back to the simple disc pattern. If we wish the casting to come out as cast to the dimensions of 12 inches in diameter and 1 inch thick, we must make the pattern larger and thicker than 12 inches × 1 inch to compensate for the amount that the metal will shrink when going from a liquid to a solid. This procedure is called *pattern shrinkage*. It varies with each type of metal and the shape of the casting.

The added dimensions are incorporated into the pattern by the pattern maker by using what is called *shrink rulers*. These rulers are made of steel and the shrinkage is compensated for by having been worked proportionally over its length. Thus a 3/16 inch-shrink rule 12 inches long will be actually 12 3/16 inches long, but will look like a standard rule. But, when layed out against a standard ruler it will project 3/16 inch past the standard ruler. These rulers come in a large variety of shrinks. Generally the shrinkage allowance for brass is 3/16 inch per foot, ⅛-inch per foot for cast iron, ¼-inch per foot for aluminum and ¼ inch for steel. This would hold true for most small to medium work. For larger work the shrinkage allowance is less, in some cases 50 percent less. Where a small steel casting in steel would require ¼ inch per foot shrinkage allowance, a very large steel casting might require only ⅛ inch per foot shrinkage

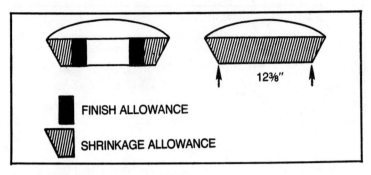

Fig. 9-2. Finish and shrinkage allowance.

allowance. So, from this we see that if we wish to cast a bar in brass 1 foot long we must make the pattern 1 foot and 3/16 inches long.

MACHINING ALLOWANCE

Now the plot thickens. Say the disc we want in brass requires that the outer diameter of the casting is to be machined (the 12-inch dimension is a machined dimension). We must then allow for machining to our 12-inch dimension. This allowance must be in addition to the shrinkage and draft allowance taken at the short side of the pattern or smallest diameter (Fig. 9-2).

We must have a pattern dimension of 12 3/16 inches. The 3/16 inch allows for shrinkage plus 1/16 inch for metal to come off. So we need an actual diameter on the small end of our pattern of 12⅜ inches.

If we dimension our layout as 12 and 1/16 inches (the 1/16 inch for machining) and we use a 3/16-inch shrink ruler to measure this dimension, then when you build the pattern it will come out fine. Or, make your pattern layout read 12 inches in diameter taking the 12-inch dimension off of a ⅜-inch shrink ruler.

Approximate finish allowances including the draft are, brass 1/16 inch, aluminum ⅛ inch, cast iron ⅛ inch, cast steel ¼ inch.

On a blue print given to the pattern maker for large projects, all finishes should be noted (turned, ground, etc.). He will then know from experience how much to allow and how much shrinkage to add to the pattern.

PRODUCTION PATTERN

In the case of the disc, if you are only going to make a casting from the pattern now and again one at a time, dimension as just described. This pattern is called a *production pattern,* one from which the actual castings are produced.

MASTER PATTERN

Now suppose we want to make one or more production patterns out of cast aluminum from which we intend to make production aluminum castings. In this case we need a wood pattern from which to cast our production pattern. If we wanted as our finished or end product a cast aluminum disc, we would have to make our wood pattern with a double aluminum shrink rule or ½ inch per foot shrinkage. As we are going to take ¼-shrinkage in going from our wood pattern to our cast pattern and another ¼ inch to our end product, these rules are called *double shrink rulers.* If we were going from a wood pattern to an aluminum production pattern to a brass casting as an end product the shrinkage allowance on the wood pattern would have to be ¼ × 3/16 or 12/16 inch plus finish, etc., if any. This type of pattern (the wood) is called a *master pattern*—a pattern from which the production pattern or patterns are made.

PARTING LINE

On the simple disc pattern of Fig. 9-1, note that the upper face of the pattern is designated as the parting line or parting face. By this we mean a line or the plane of a pattern corresponding to the

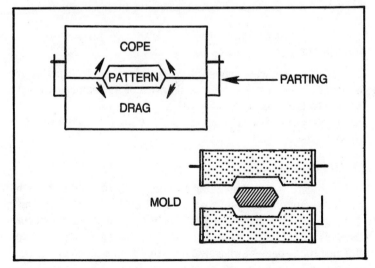

Fig. 9-3. Parting line.

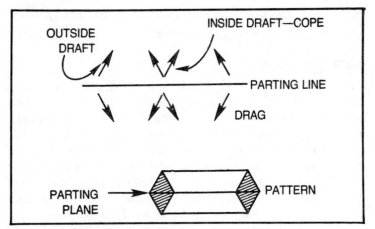

Fig. 9-4. Locating parting line.

point of separation between the cope and drag portions of a sand mold. The parting may be irregular or a plane, as the mold must be opened, the pattern removed and then closed for pouring without damage to the sand. The parting line must be located where this can be accomplished. The portion of the pattern in the cope must be drafted so the cope can be removed and the same for the drag (Fig. 9-3).

Any vertical portion of the pattern in either the cope or drag portion must be drafted or tapered as shown in Fig. 9-4. The junction or change of draft angle indicates the proper position of the parting line.

BACK DRAFT

If the pattern were shaped as in Fig. 9-5 and the mold parted at its upper face, the back draft would prevent its removal without damage to the mold.

A back draft is a reverse taper which prevents removal of a pattern from the mold.

GATED PATTERN

The gate is the channel or channels in a sand mold through which the molten metal enters the cavity left by the pattern. This channel can be made in two ways. One is by cutting the channel or channels with a gate cutter, or by the pattern having a projection attached to the pattern which will form this gate or gates during the process of ramming up the mold (Fig. 9-6).

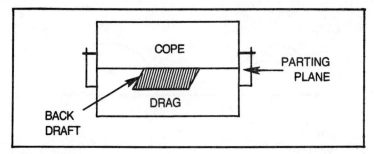

Fig. 9-5. Back draft.

If a pattern is made for a gate, but not attached to a pattern and only placed against it while making the mold, this pattern is called a *set gate pattern*.

SPLIT PATTERN

A split pattern is a pattern that is made in two halves split along the parting line. The two halves are held in register by pins called *pattern dowels*. The pattern is split to facilitate molding (Fig. 9-7).

The dowels hold the two halves of the pattern together in close accurate register, but at the same time are free enough that the two halves can be separated easily for molding like the pins and guides of the flask.

The dowels are usually installed off center in such a manner that the pattern can only be put together correctly (Fig. 9-8).

MEDIUM PATTERN

A pattern that is used only occasionally or for casting a one time piece is usually constructed as cheaply as possible. If it is a split

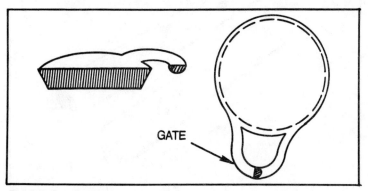

Fig. 9-6. Gated pattern.

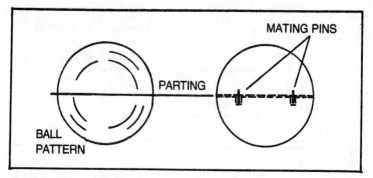

Fig. 9-7. Split pattern.

pattern wood dowels are used for pins and fit into holes drilled into the matching half. This type of pattern is called a *medium pattern*.

A core is a preformed baked sand or green sand aggregate inserted in a mold to shape the interior part of a casting which cannot be shaped by the pattern.

When a pattern requires a core, a projection must be made on the pattern. This projection forms an impression in the sand of the mold in which to locate the core and hold it during the casting. These projections are called core prints and are part of the pattern (Fig. 9-9).

Sometimes it is possible to make a pattern in such a way that a core will remain in the sand when the pattern is removed. The pattern for a simple shoring washer, illustrated in Fig. 9-10 is made in this way.

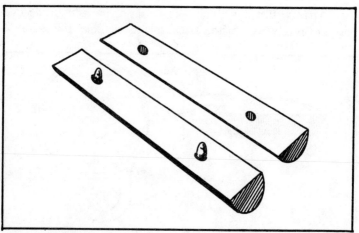

Fig. 9-8. Offset pins.

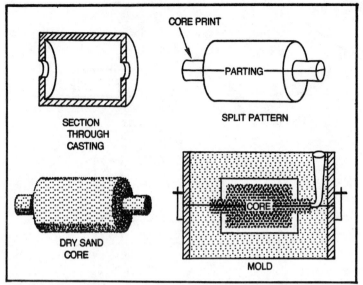

Fig. 9-9. Core prints.

MOUNTED PATTERN

When a pattern is mounted to a board to facilitate molding, it is called a mounted pattern. In this case the mount has guides on each end which match up with the flask used to make the mold. The plate is placed between the cope and drag flask, the drag rammed and

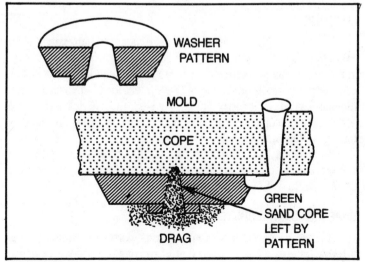

Fig. 9-10. Self-coring pattern.

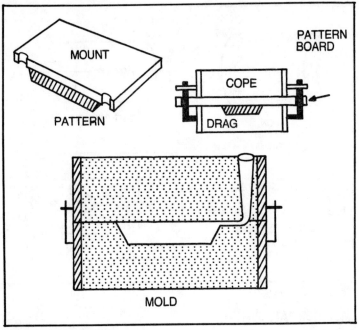

Fig. 9-11. Mounted pattern.

rolled over. The cope is then rammed and lifted off. The plate with the pattern attached is lifted off of the drag half. The mold is then finished and closed (Fig. 9-11).

MATCH PLATE

The match plate is the same as the mounted pattern with the exception that when you have part of the casting in the cope and part in the drag (split pattern), these parts are attached to the board or plate opposite each other and in the correct location so that when the plate is removed and the mold is closed, the cavities in the cope and drag match up correctly. The molding procedure is the same as for one sided mounted plate (Fig. 9-12).

In most cases all the necessary gating runners, etc., are built right on the plate. The match plate might have only one pattern or a large quanitity of small patterns.

COPE AND DRAG MOUNTS

In this case you have two separate pattern mounts. One is fitted with female guides for the drag and one is fitted with pins for the cope. These must match up with the flask used.

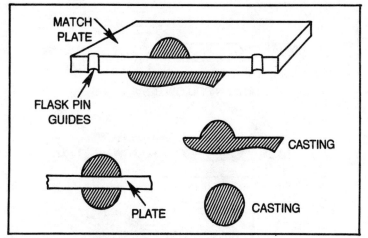

Fig. 9-12. Match plate pattern.

The cope half of the pattern is attached to the cope mount and the drag pattern is attached to the drag mount. The cope and drag molds are produced separately and put together for pouring. The usual practice is for one molder to make copes and another to make the drags. Cope and drag mounts are quite common when making large and very large castings where a match plate would be out of the question due to its bulk and weight. Cope and drag mounts are sometimes called *tubs* (Fig. 9-13).

FOLLOW BOARD

A board with a cavity or socket in it which conforms to the form of the pattern and defines the parting surface of the drag is called a

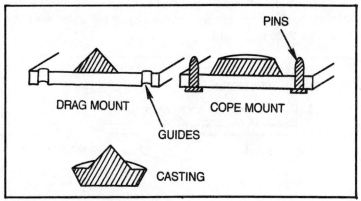

Fig. 9-13. Cope and drag mounts.

151

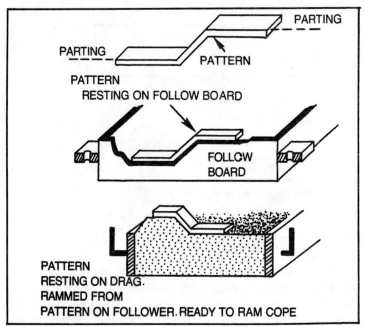

Fig. 9-14. Follow board.

follow board. It can be made of wood, plaster or metal. When made of sand it is called a *dry sand match*.

The pattern rests in the follow board while making up the drag half of the mold and in doing so establishes the correct sand parting. The follow board is removed leaving the pattern rammed in the drag up to the parting. The cope then takes the place of the follow board and is rammed in the usual manner (Fig. 9-14).

A simple follow board might consist of a molding board with a hole in it to allow the pattern to rest firmly on the board while the drag is rammed (Fig. 9-15).

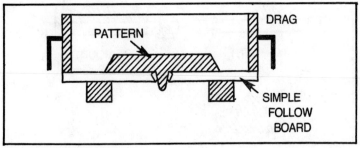

Fig. 9-15. Simple follow board.

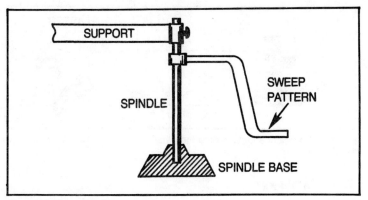

Fig. 9-16. Sweep pattern.

MISCELLANEOUS PATTERNS

When several different loosely gated patterns are assembled as a unit to be molded in the same flask, this arrangement is called a *card of patterns*.

A *sweep pattern* consists of a board having a profile of the desired mold, which when revolved around a suitable spindle or guide produces that mold. Two are usually required. One sweeps the cope profile and the other the drag profile (Fig. 9-16).

The *skeleton pattern* is a frame work of wooden bars which represent the interior and exterior form and the metal thickness of the required casting. This type of pattern is only used for huge castings (Fig. 9-17).

An *expendable pattern* is when the pattern is lost. Expendable patterns for sand casting are styrofoam. They are shaped to the desired form with attached styrofoam gates, runners and risers. The styrofoam pattern is molded with dry clay-free sharp silica sand in a box or steel frame. The pattern is vaporized by the metal poured into the mold, leaving the casting.

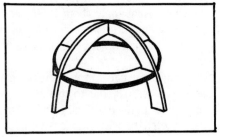

Fig. 9-17. Skeleton pattern.

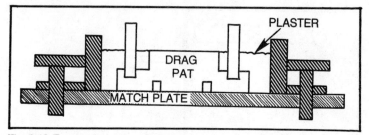

Fig. 9-18. To mount a pattern, drag the patterns, plug mounting holes and pour the plaster.

WOOD PATTERNS

Wood patterns used for casting are given several coats of orange shellac to which a pinch of oxalic acid has been added. This gives them a waterproof, smooth and hard surface.

The majority of wood patterns are made of white pine (sugar pine) as it is easily worked and when shellacked properly will not warp under ordinary use.

The approximate weight of a casting can be determined by weighing the wood pattern and multiplying by the appropriate factor indicated. Aluminum 8, cast iron 16.7, copper 19.8, brass 19.0, steel 17.0.

A white pine pattern weighing 1 pound when cast in aluminum will weigh 8 pounds, in brass, 19 pounds, etc.

MOUNTING PATTERNS

The brass plate method of transfer is an old one and, if done with enough care, usually will do a good job. It is my opinion, however, the use of a suitable transfer frame or flask and plaster is by far the easiest method of getting a close match.

This system is simple, cheap and pretty close to foolproof (Figs. 9-18 through 9-20). Take a 12 × 18-inch flask, for example, and mount patterns on a plate for use in it. Remove the bushings from both ends of the cope and rebush the flask with tight, round bushings. Allow only about 0.003-inch clearance between the pin and bushing. Alternatively, you can design, cast and machine a special aluminum flask with a close fit between the pins and the bushings.

Finish the patterns in the usual manner, pinning them together as if you were going to use them for loose molding, with one exception. In the drag-half patterns, drill suitable holes through the prints, etc., to use to mount the patterns to the plate in the final

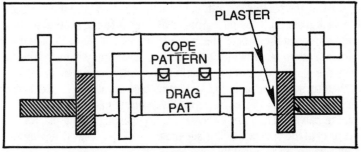

Fig. 9-19. The second step in mounting a pattern includes removing the plate, fitting the cope patterns to drag patterns and pouring the cope plaster.

steps. Next, fit a suitable 12 × 18 inch aluminum plate to the doctored-up flask with suitable pin lugs. Hold these pin lugs also to a close tolerance—0.003 inch or so clearance between the lugs and the pins.

Place the cope and drag flask together, with the plate between them. Turn the entire rig so that the drag side is up. Locate the drag patterns on the plate where you want them with respect to gating. Glue them in place with shellac or wax. Place suitable dowels in the holes you drilled for use later during mounting. These dowels should stick up out of the pattern not less than an inch or so.

With the drag patterns properly located and stuck down to the plate and with plugs or dowels in the mounting holes, place a weight on each corner of the flask. To the surface of the plate and patterns, apply a good commercial plaster release. Pour enough plaster into the drag flask to fill it to a depth which will give a good coverage around the patterns, yet leave the dowels sticking through the plaster.

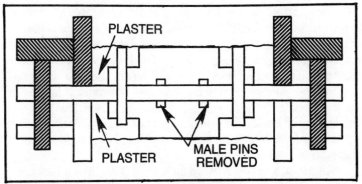

Fig. 9-20. The third pattern mounting step involves removing halves of flask containing plaster, patterns and plate plugs. Then drill the mounting holes.

After the plaster has set, remove the plate and the drag half of the flask from the cope half. Apply a little heat to the bottom of the plate to loosen the wax or shellac and remove the plate. Should the patterns come out of the plaster in the drag, simply remove them from the plate and reset them in their respective cavities. The drag is then turned over on its back, parting face up. Clean off patterns and top carefully. Place the cope halves of the patterns over their mating halves. They are imbedded to the parting line in plaster in the drag of the special flask. Place the cope on the drag, weighing all four corners with a suitable weight.

Apply a plaster parting or plaster release over the entire surface of the plaster and the cope halves of the patterns. Then pour sufficient plaster into the cope to give good coverage around the cope patterns. Since you have no dowels to worry about, you can cover the patterns up. Part the two flasks. The cope half will contain the cope halves of the patterns and the drag half will contain the drag halves.

Next, return to the aluminum match plate fitted with suitable lugs and replace the cope half. Roll the whole works over to bring the drag up and clamp all four corners. The clamp should extend from the cope or top of the flask to the bottom of the drag half. With the flask in the upside-down position, remove the plugs or dowels in the holes for mounting. Leaving these holes open, take the whole business over to the drill press and drill through the drag patterns, the plate and the cope. Chip the plaster away until you get down to the pattern. Then drive pins through or screw them down. There are many variations of this method, but this is the basic principle. Remove the special flask and the plaster, leaving the patterns with a very close register. Also remove the special lugs from the plate and mount whatever type of lugs you are using.

This rigging, done in the same manner, is suitable for making cope and drag mounts. It also will work in the mounting of a symmetrical match when the flasks are fitted up with a pin on one end and a bushing on the opposite one. This is the type of equipment used on a pin-lift or stripper. The first half off, the drag, is turned end-for-end, and the second half becomes the cope. It is used in some plumbing shops as well as in hobby shops which engage the production of small and medium castings. The plaster transfer system for mounting these symmetrical matches is used widely in this type of work.

Although a new 12 × 18-inch steel flask which has been rebushed can be used, I would recommend that you construct a special pattern-mounting frame. This frame, or pair of frames, should be machined up very accurately in a tool and die shop. Special pins and bushings can be made and you will find this equipment to be more suitable.

In most cases you require only about 2 or 2½ inches of depth per section since it is not necessary for the plaster to cover the entire pattern. Only enough plaster is necessary to grip the pattern and hold it in position until it is mounted firmly to the plate. When mounting patterns on cope and drag plates, be sure that the cope frame and drag fit accurately, that the drag frame fits the drag plate and that the cope frame fits the cope plate. In the case of cope and drag mounts with an elongated or slotted bushing on one end, the bushing must be removed and replaced with a round bushing during the pattern mounting procedure only. It later is replaced with the original bushing.

There are many variations of the plaster transfer method and with a little fiddling around you might come up with something more suitable to your particular operation than the foregoing method. However, this is the basic procedure. Whether the patterns are pinned or screwed to the plate is a matter of individual preference.

In my opinion, it is poor practice to drill patterns prior to mounting on plates, tap out the drag half or cope half of the pattern, mount the patterns by drilling oversized holes through the matchplate and rely only upon the screw which passes through the cope half of the pattern and the plate and into the drag half of the pattern.

I have had experience with clamping patterns together and with the screw method. I also have drilled through the two halves of the patterns and mounted the patterns across from each other on a match plate by driving a pin through the hole and an oversized hole in the plate and into the remaining half of the pattern. If by chance the drill does run off, however, the farther the patterns are separated, the greater the mismatch will be.

With the plaster transfer method, however, it doesn't make much difference which way the drill goes as long as the patterns are secured tightly to the plate prior to the removal of the plaster. The plaster will hold patterns accurately across from each other during the drilling, tapping or pinning operations. Any type of method for mounting can be used in conjunction with the plaster.

The accuracy of this system depends entirely upon how accurately and closely your transfer frames fit up.

RECESSED PATTERN

When molding a pattern with a deep recess in the cope side of the pattern, even if you don't come up quite straight with your draw and a section of the green sand stays down, you can often still retrieve and save the cope.

In many cases this problem is caused by improperly tempered molding sand and you will find that the sand is dry where the drop broke loose from the cope. Often the sticky pattern with low green strength and soft ramming plus the pocket is too deep or heavy to come up with the cope.

Regardless of the cause there are two ways of saving the cope. Put the cope back on and roll the mold over. Draw the drag off leaving the drop resting on its mating surface. Now push a large headed nail carefully into the sand to pin the broken piece to its mate. Turn the cope over and see if you have done a good job. If the piece that stayed down is not too large, the next best bet is to dampen both surfaces with a bulb sponge and sprinkle the surface of the sand piece in the pattern with wheat flour. Put the cope back in place and strike directly over the problem with the butt end of the rammer. Allow a second or two for your flour glue to stick, then draw the cope, roll it over and nail it. In either case be sure to patch or slick the junction of the repair.

Part 2
Dictionary

abrasives: Any material for grinding, polishing, blasting etc., which is abrasive. The field of abrasives runs from a very mild abrasive such as *rottenstone* tin oxide to highly abrasive diamond dust, or grit. Natural abrasives include talc, sand, pummice, emery, corundum, garnet, diamond, etc. The main manufactured abrasives are silicon carbide, aluminum oxide, metallic shot and grit. The forms are as many as the types of abrasives—sandpaper, grinding wheels, rouges etc.

acetylene: A colorless gas HC:CH. It is used as a fuel in high temperature torches such as the oxygen/acetylene welding and cutting torches. Pure acetylene has a sweet odor but when contaminated with hydrogen sulphide it smells quite bad. It is easily generated by the action of water on calcium carbide. It is sold in pressure tanks called bottles dissolved in acetone to render it nonexplosive. One volume of acetone will absorb or dissolve 25 volumes of acetylene. It is used in producing other chemicals. Large users often produce their own with a device called an *acetylene generator*. It consists of two chambers, one above the other. The top chamber contains dry calcium carbide. The bottom chamber contains water. The carbide is dispensed into the water by a pressure-regulated valve, or gate, when a predetermined acetylene gas pressure is reached by the reaction in the bottom chamber. The valve closes preventing any more carbide from entering the water. When the pressure drops due to the use of the acetylene from the bottom chamber, the valve opens and the process is repeated.

In Sweden during World War I, some automobiles were operated on acetylene gas. Presto-O-Lite (r) Union Carbide Co. is the trade name for acetylene dissolved in acetone.

acid: Ceramic refractory materials of a high melting point consisting largely of silica. Thus steel melted in a furnace with an acid refractory bottom under an acid siliceous slag is called *acid steel*.

adjustable jackets: A loosely cornered slip jacket which will adjust to the taper of the mold affording a good close fit (Fig. A-1). A popular brand adjustable jacket sold under the name Wopper Jaw jacket is manufactured by Products Engineering, Cape Girardeau, MO.

Fig. A-1. An adjustable jacket.

aerators: Any device which fluffs up and introduces air into a molding sand which increases the flowability and rammability of the sand (Fig. A-2). It also helps to evenly cool and distribute the moisture. Simple riddling of the sand accomplishes this evenness to some extent. Some machines condition and aerate the sand in the same action. Such a device widely used in the foundry is the Royer Sand & Blender machine. This machine separates, aerates and blends all in one action. Royer also produces many types of aerators and conditioners.

Most mullers have an aerator at the discharge snoot which consists of a series of knives on a high speed shaft. The sand being discharged must make a trip through these whirling knives which aerate the sand. Some machines consist of steel brushes rotating at high speed through which the sand must pass. The process is called *aerating*.

aerators for foundry sand: Originally patented in 1910, the cleated-belt aerator was introduced to the foundry industry by Royer and was quickly accepted for conditioning sand. Over the years, the design, construction and materials of the unique

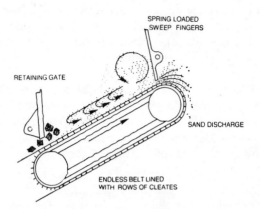

Fig. A-2. An aerator. The refuse falls at the retaining gate.

machine have been continually improved. The basic aerator has been adapted to meet the industry's ever-increasing requirements for proper sand preparation. The development of new aerators for foundry applications continues to be an on-going process. Royer manufactures a variety of standard aerators: basic stationary and portable units, with or without magnetic cleaning; aerators mounted in bucket elevators; belt-mounted aerators; aerators for feeding automatic molding machines; and aerators on molder's hoppers.

As sand is charged into a Royer Aerator, it falls on an inclined, endless belt which has rows of closely-spaced alloy-steel cleats. With the belt moving at more than 100 fpm, the cleats tear apart lumps and propel the sand up the belt to a *sweep*—a row of flat spring-loaded fingers mounted above the belt at the head pulley. As the sand churns against the sweep, trapped gases are released and foreign materials, such as wedges and cores, fall to the lower end of the belt. The tumbling-churning-mixing action on the belt aerates and cools the sand. Granular sand, between the rows of cleats, passes under the sweep and is thrown forward. The stream of free-flowing sand continues to cool as it is discharged into the air. This is called the basic cleat-belted principle.

The pattern of cleats, speed and inclination of the belt and spring tension are basic design parameters. The clearance between the sweep assembly and the belt, however, can be adjusted to achieve the desired degree of aeration. Capacity is determined by the width and length of the belt and the condition of the sand.

163

A.F.S. clay: American Founderman's Association defines clay as the earthy portion of a foundry sand which when suspended in water fails to settle inch per minute and consists of particles less than 20 microns (.008 inch) in diameter—a combination of true clay and silt.

A.F.S grain distribution: The distribution in per cents of various sizes of grains in a given sand established by running a sample through a standard set of screens and weighing the sand retained on each screen. The distribution is then expressed on a graph (Fig. A-3). Sand with 70 to 80 percent retained on three adjacent screens is considered ideal.

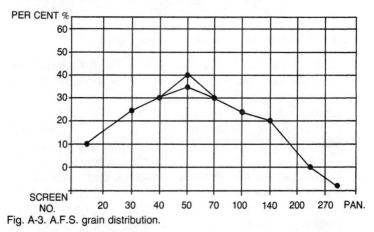

Fig. A-3. A.F.S. grain distribution.

age hardening: The property of some ferrous and nonferrous alloys to increase hardness and strength by time alone at room temperature. Tenzaloy(TM) Federated metals is a popular aluminum casting ingot which has very good properties of age hardening. The normal composition is copper 0.6 percent, zinc 7.5 percent, magnesium 0.4 percent and the remainder aluminum. Cast aged one day at room temperature has a tensile strength of KSI 29. After 10 to 14 days the tensile jumps to KSI 35. The Brinell hardness (500 Kg) jumps from 60 on the first day to 74 about two weeks later. Permanent mold cast Tenzaloy tensile goes from 30 to 40. Some soft metals (lead base) go the other way losing strength with age.

air belt: The belt surrounding the cupola or blast furnace which receives the air from the blower and evenly distributes it to the *tuyeres.* Also called *bustle pipe.*

air dried: A green mold or core which is allowed to dry naturally. Some bonding clays used in green sand molding gain strength after the mold has air dried. This increase in strength is called *air set strength*. Holloysite clay from Utah has this property due to its lath-shaped structure. Also known as *white clay*.

air floaters: Any dry compound that is air floated or capable of floating in the air—fine dust, etc. Compounds of pitch, dry parting, blacking, clays and flours that are advertised as floaters or air floaters indicate that they are free from any heavy particles. They are lightweight, usually of a fine uniform size free from tramp material.

air furnace: A reverberatory type of furnace. The metal is melted on a shallow hearth by the flame from the fuel burning on one end of the hearth passing over the bath on its way to the stack at the other end. The heat is reflected from the roof and sidewalls. Air furnaces are fired with natural gas, oil and pulverized coal or coke. The passive air furnace is fired by a natural draft pulling the products of combustion across the hearth. A shallow bath is necessary in order to bring the bath temperature up high enough. A deep bath will not melt properly due to heat convection through the hearth bottom.

The air furnace is widely used in the production of malleable iron. It is hard to control when using the cupola to produce malleable by itself, but is a fast and efficient method of melting. Very often the metal is melted in a cupola and then tapped into the air furnace for refining to the correct analysis. This process is referred to as *duplexing*. The air furnace is a close cousin to the *open hearth* furnace.

With the older, more passive air furnaces which used coal with the fuel burning on a grate, it was quite common to use a blower for forced draft along with the *stack draft* (suction). A pulverized coal burner has taken the place of these old jobs. A small air furnace burning coal the old way with some air assistance is a great melter for the backwoods or hobby foundry for ferrous or nonferrous melting. The air furnace is usually provided with a breast and tap hole on each side and is tapped like the cupola. As you have a rather large metal surface exposed to the products of combustion (unlike a crucible), great care must be exercised as to the conditions of combustion or you can wind up with a highly oxidized melt. This is true of any open hearth air furnace of the small ferrous type and of the nonferrous reverberatory furnaces which have come into popularity.

air hardening: When certain alloys of steel are cooled in air from a temperature above or higher than the transformation range to room temperature and remain hard, they are known as *air-hardening alloys* or *self-hardening steels*. Metal lathe tool bits are self-hardening steel. They are extremely difficult or impossible to anneal.

air hole: Gas holes. Air or gas trapped in the casting during solidification. The major cause of *gas porosity* is gas absorption by the metal from the products of combustion, sulphur dioxide, hydrogen and oxygen. The least likely cause is a too hot pouring temperature.

airless blasting: The cleaning of castings by throwing shot or steel grit against the objects to be cleaned, Similar to sand blasting (Fig. A-4). The basic difference is that air is not used to propel the abrasive. The abrasive is thrown by a high-speed

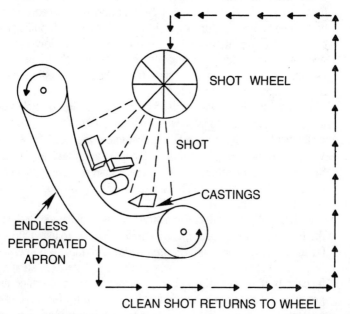

SHOT WHEEL

SHOT

CASTINGS

ENDLESS
PERFORATED
APRON

CLEAN SHOT RETURNS TO WHEEL

Fig. A-4. Airless blasting.

impeller wheel. The shot or abrasive is fed through the hub of the impeller wheel where it is picked up by the throwing blades. It is accelerated as it passes to the periphery of the wheel and is thrown at high velocity. There are variations of the machines but the principle is the same. Some are continuous where the castings are carried through the blast chamber in one side and out the other. Some of the castings are placed on rotary tables which rotate in the chamber and others are rolled back and forth on an endless apron conveyor (metal or rubber) which tumbles the castings under the blast of abrasive. The abrasive falls through the perforations in the apron and is conveyed back to the wheel via a screw and bucket. Along the way back to the wheel, the shot or grit is cleaned of any sand, dust, etc.

air pocket (permanent mold): A defect known as an air pocket caused by trapped air or gas in the mold cavity. Found with permanent molds as well as with die casting. First it has to be determined if the pocket is caused by a low pouring temperature; by pouring incorrectly (too slow, bubbling etc.); by the gating system not being properly designed; or by the mold not running hot enough. When you have eliminated all of these possibilities, then the mold must be vented.

air rammers: A rammer operated by air introduced into a cylinder which drives a reciprocating piston and rod (Fig. A-5). A rubber butt or peen is attached which produces a fast rapid blow. Widely used for bench, floor molds and in the ramming of cores. They range from small bench rammers to heavy floor rammers.

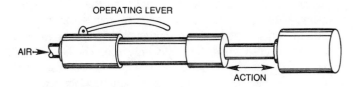

Fig. A-5. Air rammer.

air strength: When a green sand specimen is allowed to stand for any length of time and air dry before it is tested for its compressive strength, the air strength will have increased. The difference between a sample tested before and after air drying gives the foundryman an idea of what happens when a mold sits for a length of time and air dries prior to pouring. The increase in strength varies widely depending on the type and kind of clay used as a bond. Most foundrymen pay little attention to the air strength of their molding sand and are puzzled about dirt inclusions in the casting which they attribute dirty molding. The problem could be simply that the sand is too low in air strength. During air drying, sharp edges lose strength and crumble off or wash off into the mold cavity. A simple change in clay bond can prevent the problem. The moisture content also plays a part in the loss or gain in air strength of the sand.

air vibrator: Any pneumatic type of vibrator (Fig. A-6). There are two basic operating principles. One is the free piston type where when air is introduced into the cylinder, the piston shuttles back and forth at high speed (a floating piston air engine). The other is the ball type where a steel ball travelling in a case produces the vibration. The most common type used in the foundry is the *match plate air* (piston) vibrator. It vibrates the pattern to insure a clean lift of the cope and the pattern.

MATCH PLATE VIBRATOR

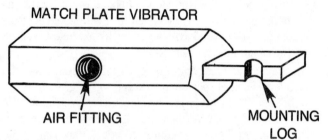

AIR FITTING   MOUNTING LOG

Fig. A-6. Air vibrator.

Albany sand: A naturally bonded sand from the district around Albany, New York. This sand was strip mined for years and years in various grades from sand suitable for light aluminum casting to course grades for heavy gray iron. The field is pretty well worked out of good quality grades. Synthetic molding sands have just about replaced the use of naturally bonded sands. *Naturally bonded sands,* however, can be found practically anywhere in the United States. They contain sufficient clay bond and the correct properties needed to make molds as mined.

alkali metals: 1A of the periodic system including lithium, sodium, potassium, rubidium, cesium and francium.

alkaline metals: Calcium, strontium, barium, radium and the group 11A of the periodic system.

allotropy: The occurrence of an element in two or more forms such as carbon which occurs in nature as diamonds, soft graphite and amorphous coal.

alloy: A combination of metals melted together. If you take 85 pounds of copper, 5 pounds of tin, 5 pounds of lead and 5 pounds of zinc and melt them together, you'll have an alloy known as *red brass*. Also called *ounce metal*, it is an old and often used casting alloy. With 50 percent lead and 50 percent tin melted together you have the common soft solder known as 50/50. The percentages of the alloying elements vary giving you an alloy of a different name and characteristics as to castability, melting point, etc. If you took the amount of lead and tin just described (solder) and raised the percent of tin to 80 percent and lowered the lead to 20 percent, you would have an alloy called *babbit*, a bearing metal. Raise the tin even higher to 90 to 95 percent and lower the lead to 10 or 4 percent and you'll have *pewter*. Some metals or constituents are not compatible in any amounts other than a trace. The trace would be called an *impurity*. From extensive metallurgical and physical testing, the amounts of various metals in percents that will successfully alloy with each other have been fairly well established. The development of new, useful, and sometimes sophisticated alloys are being researched and developed daily. For example, silicon bronze, which is an alloy of basically 3 to 4 percent silicon and the remainder copper is not compatible with lead other than a very minor trace which would be classified as a harmful impurity. Lead in even small amounts in a silicon bronze combines with the silicon to form lead silicate (glass). When some element is added to an alloy in a minor amount to change the characteristics of the alloy being produced, it is referred to as a *hardener*. Most babbit metal contains a small percent of antimony which increases the hardness of the alloy considerably.

alloy cast iron: Cast iron is composed of iron, carbon silicon, phosphorous, manganese and sulphur. The elements which compose the alloy called cast iron or *gray iron* are used in different and varying proportions depending on the grade of casting desired. Typical common gray iron would be approximately 93 percent iron, 3.25 percent carbon, 2.5 percent silicon, .50 percent manganese, .65 percent phosphorous and .10 percent sulphur. Sulphur is usually considered an impurity.

Alloy cast irons are essentially an alloy or iron and carbon to which an element or elements in various percentages and combinations are added. Sufficient amounts produce a measurable modification of the physical properties of the iron in the section (thickness & weight) under consideration. *Ferro alloys* such as ferrosilicon, and the nonferrous group phos copper, copper nickel, etc., are called *alloying elements*.

Some of these additives include chromium, silicon, molbydenium, vanadium, nickel, titanium and phosphorous copper. These additions are usually added in the form of ferro alloys purchased by the percent produced by smelting companies. More often than not the elements by themselves are difficult if not impossible to add as is from the physical element as well as the cost due to loss by oxidation. These additives are sold combined with iron in the form of shot wafer briquettes as to the percent of the element combined with the iron. Thus if you purchased 100 pounds of 50 percent ferrosilicon you would have 50 pounds of silicon and 50 pounds of iron for each 100 pounds. When melting with the cupola, these additives are usually added to the stream of metal coming from the spout to the receiving ladle or to the pouring ladle, otherwise most of it would be lost if added to the charge in the cupola. In reverberatory, crucible or induction melting they can be added directly to the bath. The term *innoculance* is used quite often for these additions.

alpha iron: The form of iron characterized by a body-centered cubic crystal structure that is stable below 1670°F. Carbon is practically insoluble in alpha iron when heated to a temperature above 1670°F. The *ferrite* changes from alpha to gamma state, a face-centered cubic form. *Gamma iron* readily dissolves carbon. The solid solution is known as *austenite*.

alumel: A nickel-based alloy used chiefly as a component in thermocouples.

aluminite: An alumina refractory material.

aluminum: A light metal produced from boxite which is not sufficiently strong enough to be used to any extent commercially. It is usually alloyed with silicon, copper, magnesium, nickel and zinc in various amounts and combinations to produce various physical characteristics as to castability, strength, soundness, etc., depending upon the method chosen to cast, the castings and use. There are a great number of alloys available (Table A-1).

aluminum bronze: A copper base alloy containing 5 to 15 percent aluminum and up to 10 percent iron, with or without manganese or nickel (Fig. A-7).

- Normal composition—88 percent copper, 3 percent iron, 9 percent aluminum.
- .276 pounds per cubic inch.
- Patternmakers shrinkage ¼ inch per foot.
- Pouring Temperature: Light castings 2050–2250°F., heavy castings 200–2100°F.
- Tensile strength, KSI 80.

Aluminum bronze is a problem metal to cast due to its high shrinkage, short freezing range and the tendency to form *dross* that must be trapped and prevented from getting into the casting. The same precaution on gating applies to both small and large

Sand Casting Alloys		
Copper 1.8% Manganese 1% Aluminum Remainder Copper 4% Silicon 3% Aluminum 93% Copper 4% Silicon 5% Aluminum 88% Copper 10% Magnesium .3% Aluminum 89.7%	Silicon 5% Aluminum 95% Copper 4.5% Silicon 5.5% Aluminum 90% Copper 7% Silicon 3.5% Aluminum 89.5% Copper 10% Silicon 4% Magnesium .3% Aluminum 89.7%	Copper 4% Silicon 1% Aluminum 95% Copper 3.5% Silicon 6.5% Aluminum 90% Copper 6.5 Silicon 5.5% Magnesium .3% Aluminum 87.7%
Permanent Mold AL Alloyus		
Silicon 5% Aluminum 95% Copper 7% Silicon 5% Magnesium .4% Aluminum 87.64	Copper 3% Silicon 6% Aluminum 91% Copper 10% Magnesium .3% Aluminum 89.7%	Copper 4% Silicon 3% Aluminum 93%
Die Casting AL Alloys		
Copper 4% Silicon 8% Aluminum 874	Silicon 5% Aluminum 95%	Silicon 9.5% Magnesium .5% Alumimum 90%

castings. The best method of *gating* is to introduce the metal into the bottom of the mold through an inverted *horn gate* (large end attached to casting), using a *skimmer sprue* between the pouring sprue and the small end of the horn.

Risers should be one and one half times larger than the secton they are to feed. Use large fillets and avoid sharp edges and a gating system which would produce turbulence or a *nozzling effect*.

Keep the value of the moisture in the sand as low as possible. The top value should be 6 percent. Large castings should be cast in dry sand molds or no-bake molds. Any cores should be open and free from gas-producing ingredients. Keep flour to a minimum. Use a simple oil sand core. Suitable green molding sand should be somewhere within these limits: permeability, 15 to 20; claybond, 10 to 20; green strength, 5 to 10; moisture, 3 to 6 percent; and sand fineness, 100 to 150.

More often than not the weight of the risers will be equal to or exceed the weight of the casting. Heavy sections which cannot be readily fed should be chilled with adequate *chills*.

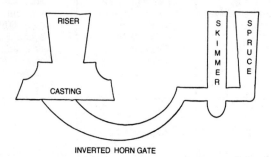

Fig. A-7. Aluminum bronze.

aluminum flux: Any material such as dry nitrogen or other insert gas bubbled through the molten aluminum to rid it of absorbed gasses or stable metallic salts such as *zinc chloride* or *aluminum chloride*. In some cases a mixture of one or more *chloride, fluoride* or *cryolite* with aluminum magnesium alloys. One or more alkaline earth salts are used as a flux. With clean metal and good melting practices you can produce satisfactory castings without the use of a flux. Consult your supplier for the recommended flux for the alloy you are melting and the type of melting equipment you are using. The simplest and safest way which produces no obnoxious poison gas is to simply bubble dri-nitrogen gas through the molten bath.

aluminum nickel bronze: When nickel is added to aluminum bronze in the order of from 2 to 5 percent, you gain in strength, reliability and corrosion resistance. Excellent stability in structure during cooling from the casting temperature prevents the *self annealing* that some bronzes suffer from if allowed to cool too slowly. It can cause deterioration in mechanical properties.

Typical aluminum nickel bronze is composed of copper, 83.0 percent; nickel, 2.5 percent maximum; aluminum, 10.0 to 11.5 percent; manganese, .5 percent maximum; and iron, 3.0 to 5.0 percent.

Aluminum Ni bronze is handled the same way you handle straight aluminum bronze as far as foundry practice is concerned with the exception that the higher the percent of nickel, the hotter you must melt and pour. For example, with 12 percent nickel your

Table A-2. Aluminum Solder Components.

Soluminium (German) AL Solder	Mourays
55% Tin	80 to 90% Zinc
33% Zinc	3 to 8 Copper
11% aluminum	6 to 12% Aluminum
Bureau of Standards Aluminum Solder	**Alcoa**
87% tin	95% Zinc
8% zinc	5% Aluminum
5% Aluminum	

Aluminum to Aluminum
91% Tin
9% Zinc

pouring range jumps from 2350° to 2550°F which is from 50 to 250° above the freezing point with this much nickel. A common error with all nickel alloys is pouring too cold.

aluminum pattern plate: An aluminum plate for mounting wood or metal patterns *(match plate)* for molding. Plates are produced in a wide variety of sizes, thicknesses and types. The common type is a flat parallel ground plate with ears to handle the plate and attach the lugs (guides) and match plate vibrator (rapper).

aluminum solder: For the various components and percentages of aluminum solder, see Table A-2.

ammonium chloride: NH4CL called *sal ammoniac.* Used as a soldering flux for soft solders, a lead flux and in brick form to clean soldering coppers and irons.

anchor (core): Any device used to hold or anchor a core in place (Fig. A-8). It could be a nail running through a print into the sand

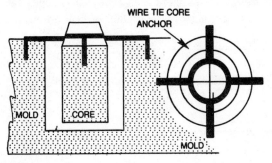

Fig. A-8. Anchor core.

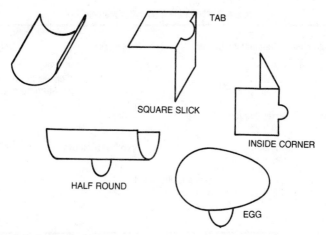

SQUARE SLICK

TAB

INSIDE CORNER

HALF ROUND

EGG

Fig. A-9. Angle sleeker.

to anchor the core or a wire tie used when pouring an open-faced bushing mold. The anchor holds the bushing core down and centered when pouring without a cope.

angle sleeker: A molder's hand tool for smoothing or sleeking inside and outside corners of a mold (Fig. A-9). These sleeker tools have a half-round tab on their back surface which is held between the thumb and first finger and used like you would a pressing iron to smooth the area you desire, especially when repairing a mold surface. Avoid excess sleeking using too much pressure. This could compact the sand at the point of sleeking too tight and result in a kick or blow at that point.

angle stem chaplet: A sheet metal chaplet with an open design to permit an adequate flow of molten metal through the chaplet (Fig. A-10). Insures proper burning in.

annealing: When most ferrous metals are heated above the *critical temperature*—temperatures at which changes take place. They are determined by liberation of heat when the metal is

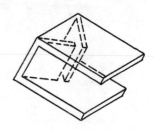

Fig. A-10. Angle stem chaplet.

heated, thus resulting in halts and arrests on heating or cooling curves. The metal is held at the critical temperature until the transformation is complete. The metal is cooled slowly in a furnace or packed in bone ash or cinders to cool slowly. The metal is then relieved of design and casting stresses and is more ductile and easier to machine.

With copper base and precious metals (gold, silver, etc.), the process is reversed. In this case the metal is heated to its critical temperature—usually a dull red in the dark—and quenched in cold water.

anneal, malleable: This is the heat treatment involving heating and then slow cooling. With regard to malleable iron, it is the method used to convert the hard brittle iron castings into tough ductile, malleable iron. This treatment causes the carbon in the white iron to precipitate in the free state in the form of tiny particles instead of flakes as in gray iron.

annealing pots: Iron pots in which castings which are to be annealed or malleabilized are packed to protect them from the furnace atmosphere during the annealing process.

antimony oxide: An oxide of antimony. Its principal foundry use is as a hardener in lead base alloys.

anti-piping compounds: Carbonous (slow oxidizing) compounds which are sprinkled on the top of the riser to slow down the heat loss, keeping the riser liquid longer so that it can do its job (Fig. A-11). Some are mildly exothermic or quite volatile like the old hot patch. *Insulating sleeves* of plaster of paris, or other insulat-

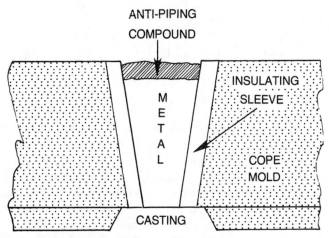

Fig. A-11. Anti-piping compounds.

ing materials are sold by foundry suppliers ina wide variety of sizes and materials. Some are simply insulators, others are *exothermic*. Exothermic anti-piping compound materials are composed chiefly of what is known as *thermit*, but modified so as to react less violently. Thermit is basically a mixture of powdered aluminum and iron oxide.

The idea is to keep the riser liquid as long as possible. It insures the proper feeding of liquid metal to the shrinking casting to promote directional feeding.

Anti-piping compounds should not be added until the riser will take no more metal. They prevent the compounds from being fed into the casting. Anti-piping compounds and sleeves are widely used in steel and iron founding and to a small extent in large nonferrous work.

arbor: A metal (cast or fabricated) form used to support a dry sand, green sand core or mold body for strength (Fig. A-12). Also used extensively in heavy loam molding. Sometimes called a *crab*.

Arbors are usually made to fit a particular job. A common use is in the casting of gray iron soil pipe fittings using green sand cores.

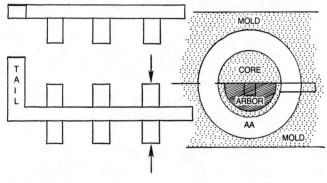

Fig. A-12. Arbor.

In practice the core is made in a full core box by *booking*. The drag half of the core box is partially filled with green sand and the arbor placed in place, then the half core completed. The cope half is then made up and the box booked. The cope box is removed and the core lifted out by the arbor which supports it. The cope half of the green sand rests on top of the arbor and the drag half supported on the arbor by the half-round sections extends into its body. The tip-tail fits into a print made in the drag mold by a print

176

on the pattern. The tip tail does two things: It prevents the core from tipping over and touching in the drag and locates the core properly in the mold when the mold is closed.

Cores for large cylindrical castings molded horizontally are often swept up (or rammed) on a perforated steel or cast iron core arbor. The arbor is first wrapped with burlap to prevent the sand from going through the vent holes (perforations).

The pattern usually has prints on it to accommodate the arbor as well as the core. Small hollow tin tubes are also sold to strengthen a core section by ramming them up where needed when making the core.

In comparison to the sculptured armature, a core or mold arbor would represent extra support and strength. In casting, it often serves several purposes all at the same time, such as a path for the core gases to escape through plus the conservation of material. In investment casting (full mold), steel tubing preforated and wrapped with paper is used as a *core support* and *core vent*. More often than not it dispells the difference from losing a casting from a broken core and a core blowing. The unperforated section extending through a section of the mold which will fill with metal when poured prevents the vent from stopping off and causing a blow. Steel automobile gas line tubing is excellent for this purpose.

arc furnaces: There are two basic types in use—the *direct* and *indirect arc* (Fig. A-13). In the direct arc furnace three electrodes in a triangular pattern come through the roof of the furnace through suitable openings. They are provided with a suitable mechanism for raising and lowering them. The metal is melted by direct contact with the electrodes. The electrodes are raised and the charge placed on the hearth. The electrodes are then lowered and the current is turned on. The electrodes melt their way through the charge nearly to the bottom. As the molten metal (bath) rises, the electrodes are raised to the correct height in respect to the bath. As soon as the correct refining period is reached, the current is reduced. The bath is refined and chemically adjusted in the furnace by various slag treatments (boils), oxygen lancings, ferro alloy additions (depending on what is called for as the final analysis) and the starting charge. There are furnaces (direct arc) as small as 2 tons to 150 tons and larger. The furnace is tilted to remove the *heat* (molten metal).

In the indirect arc furnace the metal is melted by heat radiation of an arc struck and maintained between two carbon electrodes.

The arc is maintained by an automatic screw feed which advances the electrodes to maintain the correct arc gap as the electrodes are consumed. The shell is barrel shaped and rides on a set of rollers and is rocked back and forward through an arc of about 180°. This rocking motion performs the same function as a rotary furnace by carrying the molten metal over the heated lining in all directions, does the melting. These furnaces are called indirect arc rotating furnaces and *Detroit rockers*. The arc does not come in contact with the charge.

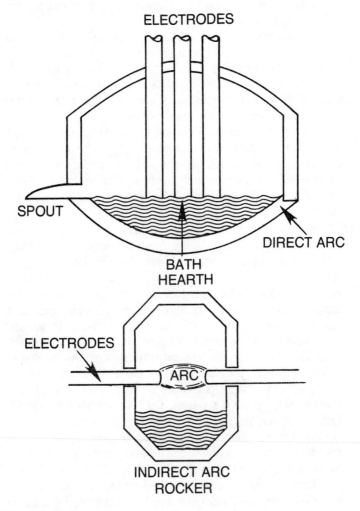

Fig. A-13. Arc furnaces.

The major manufacturer of this type of furnace is Kuhlman Electric, Detroit. The Detroit Rocker is a popular nonferrous melting unit.

arch brick: Tapered refractory bricks used to construct a furnace or kiln-arched roof. The two most common sizes are the Number 1 and Number 2 arch. By selecting the correct com-

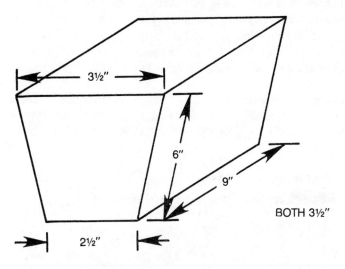

BOTH 3½"

NUMBER 1

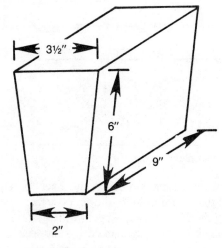

NUMBER 2

Fig. A-14. Arc bricks Number 1 and Number 2.

bination of straight brick, splits and Number 1 and Number 2 arch brick, you can build the desired arch radius.

arsenic trioxide: AS_2O_3, also known as *white arsenic*. Used as an alloying ingredient in producing steel and added to antimonial leads, alloys and white metal bearings. Used as a hardener and to increase fluidity. When added to copper, it increases the annealing temperature for use as radiators. When added to lead it changes the surface tension and permits formation of perfectly spherical shot when dropped from a shot tower. It is usually colored pink to identify it as a poison.

artificial sand: Silica sand made by grinding up silical rock and grading it.

atmospheric feeder: The theory behind the atmospheric feeder is that upon solidifying and shrinking, the casting forms a partial vacuum in the mold cavity (Fig. A-15). If a connection can be

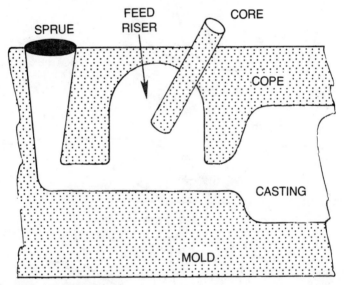

Fig. A-15. Atmospheric feeder.

made with the liquid metal in the riser (during this time and action) to the atmosphere, the *atmospheric pressure* will assist in feeding the casting and prevent *shrinkage cavities* due to lack of feed metal. It is a function of differential pressures. By placing a dry sand core extending into a feed riser or gate, you prevent the solidifying skin from closing the liquid metal off from the atmosphere allowing the atmosphere to exert its pressure to assist in feeding the casting. This system is widely used in steel castings and the author has had great success with large manganese bronze castings using this method.

This method is the principle behind *churning* a riser with an iron bar to affect feed. The object is to keep the riser open to the atmospheric pressure as long as possible. This system has been the subject of several patents by the late George Batty, Henry Phillips and John Williams.

auto oxidation: The principle behind the oil oxygen process of core making. It is a form of oxidation by which drying oils absorb oxygen and polymerize the oil to a solid. The process is accelerated by heat and oxygen bearing materials—perborates, per carbonates and peroxides.

autogenous heat treating: Shaking out the casting as soon as it has solidified, scraping off the sand and letting it cool on the foundry floor in a draft. The cooling can be speeded up with a water spray.

autogenous welding: Pouring hot metal in a defect in a casting in hopes it will weld and become part of the parent metal. This is also called *burning in*. It is a hit-and-miss deal at any rate. The hotter the casting being repaired the better chance you have of establishing a good homogenous weld.

B

babbitt: The original babbitt named after its inventor was 88.9 percent tin, 7.4 percent antimony and 3.7 percent copper with a melting temperature of 462°F. Used for machinery bearings. However, today there is a large family of tin base alloys called babbitt. Any white alloy used for bearings is currently called babbitt.

babbitt anchors: A tin form filled with baked core sand used to cast an undercut cavity in a casting where a babbitt bearing is going to be poured (cast) (Fig. B-1). The cavity made in the casting provides a lock to hold the babbit in position. The anchors have metal sprigs to hold them in place in the green sand mold during casting. They can be purchased in diameters from ⅜ inches to 1¼ inches and rectangular from ½ inch × 1¼ inches × 3/16 inch deep to ⅞ inch × 3½ inches × ⅜ inch deep.

back draft: A reverse taper on the vertical surface of a pattern which prevents the removal of the pattern from the mold.

backing board: A second bottom board used to lay the cope over on its back for removing a loose pattern, repair, etc.

backing sand: Foundry floor or system sand used to back or complete a mold after the pattern is covered with a facing sand—new or finer sand, or a facing mix such as #10 facing. (One part sea coal, 10 parts new molding sand with or without an added binder such as *goulac*, wheat flour, etc.) In order to conserve sand and materials on large molds when molding with

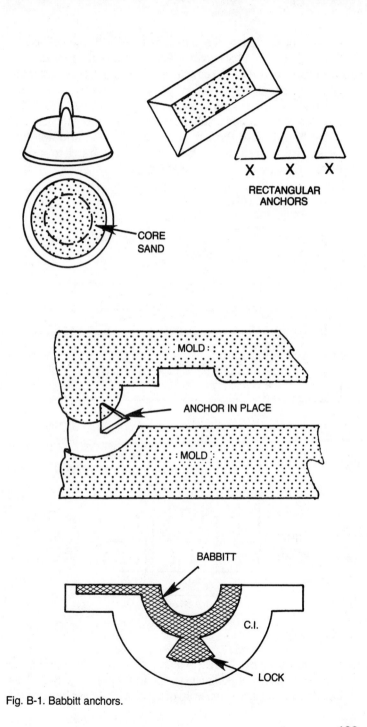

CORE
SAND

RECTANGULAR
ANCHORS

MOLD

ANCHOR IN PLACE

MOLD

BABBITT

C.I.

LOCK

Fig. B-1. Babbitt anchors.

183

no-bake or CO^2 sand, the pattern is faced with the no-bake or CO^2 and set, then backed or finished off with floor sand.

baffle plate furnace: A refractory wall in a fire box or furnace to deflect or change the direction of the flame (Fig. B-2). It is also called a *bag wall* or arch baffle. In a reverberatory furnace it is installed at the far end just before the stack causing the flame to swing back toward the burner or reverberate. In down-draft kilns it directs the flame and heat upward into the arch to prevent it from heading straight out of the stack.

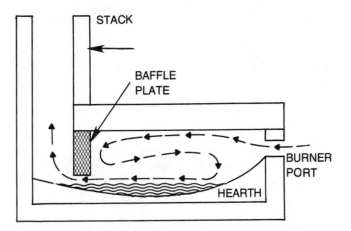

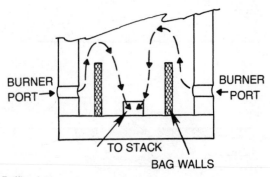

Fig. B-2. Baffle plate.

baffle plate pouring basin: The reference here to a baffle plate is in a special type of skimmer type pouring basin (Fig. B-3). This type of runner basin is made up of a cast iron runner box divided into four compartments. The theory is that each compartment affords the slag, dirt and scum to float to the top of the metal giving the metal four chances of ridding itself of foreign inclusions. When it reaches the sprue, you have only clean metal. The author has constructed various versions of this method based on the skimmer on a continuous tapped cupola spout with great success, especially in pouring large gray iron and bronze castings. The cast iron runner box is lined with loam or monolithic refractor material, blacked and thoroughly dried. The time it takes to make this type of rig is well worth it.

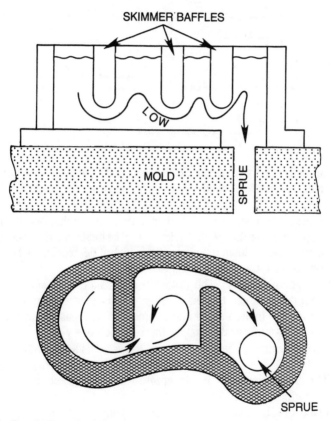

Fig. B-3. Baffle plate pouring basin.

bag house: A large chamber for holding bags used in the filtration of gasses from a furnace to recover metal oxides, fly ash and other solids suspended in the gasses. In reality it is a large vacuum cleaner. The bags are made of various materials from cloth to hi-temperature ceramic fibers. Aside from the recovery of metal oxides which might have a value, the EPA requires a bag house on many operations as air pollution devices. Bag houses are used alone or in conjunction with electrical precipitators and various washing devices. The exhaust gasses usually have to be cooled down quite a bit prior to entering the bag house to prevent the bags from melting or burning up.

bail: The hoop connection between a ladle and crane hook (Fig. B-4).

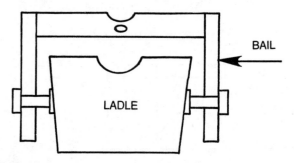

Fig. B-4. A bail.

baked core: A dry sand core that has been baked in order to set the binder. A simple oil sand core must be baked at 350° to 400°F. in order to oxidize the oil thus cementing the grains of sand together. The core stays together during the casting operation long enough for the metal to take a set. The heat from the cooling casting breaks down the binder (burns it up) destroying its bond and allowing the core to collapse.

baked permeability: The property of a baked core cooled to room temperature to pass through its gasses resulting during pouring the molten metal into the mold. These gasses must be vented away from the casting through the core vents.

baked strength: The compressive, sheer, tensile strength of a baked core or mold. A measure of its tenacity to a breaking force. The baked strength is far greater then its green unbaked strength.

baking cycle: The time required to bake an oil sand core. Minimum baking cycle in a conventional type oven ranges from a minimum of two hours to four to six hours. This will vary with the mix, size and type of core being baked. The cooling time should also be taken into consideration when planning a schedule in order to have cores to the molders on time. Everything must be considered—inspection, gluing, mudding, sizing, etc.

balance core: A core with only one point of suspension or print (Fig. B-5). The core print is made long enough to have sufficient purchase in the mold as to prevent it from moving during casting.

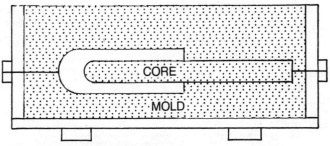

Fig. B-5. A balance core.

ball mill: A cylindrical device that is revolved. It is partly filled with iron or ceramic (porcelain) balls for grinding or mixing sand or other materials. Widely used in hard-rock mining to reduce the ore-carrying materials to a fine mesh to facilitate the removal of the ore. Ball-mill interiors are often lined with various abrasive resistant materials.

bank sand: Sedimentary sand deposits containing less than 5 percent clay.

base block: Also known as a *pedestal block*, the base block is used to support the crucible in the furnace (Fig. B-6). The base block should be made of the same material as the crucible and be the same diameter as the bottom of the crucible it supports—9 inches in height or of such height that the centerline of the burner is at the junction of the crucible and block.

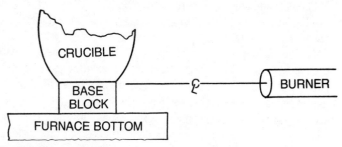

Fig. B-6. The base block.

base exchange of clay: Clay type binders have the ability to exchange some of their ions with other chemicals such as methylene blue. When the base exchange of a given clay is decreased so is its bonding power. Baroid Chemicals Inc. sells a kit used to determine the amount of effective sand binder in molding sands by methylene blue.

basic: A chemical term referring to any material which gives an alkaline reaction. In a melting furnace with the innerlining and bottom composed of a basic material such as crushed-burned *dolomite, magnesite, magnesite bricks* and using a basic slag, the reaction is called *basic melting*.

basic flask: The top frame of the flask is called the *cope* and the bottom frame is called the *drag* (Fig. B-7). In some molding operations you need one or more sections between the cope and the drag. These sections are called *cheeks*.

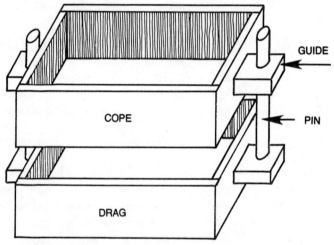

Fig. B-7. Basic flask.

The pins and guides that are used to hold the sections together can be purchased in a wide variety of types and configurations, round and half round, double round and vee shaped. Single, double or triple vee shapes can be combined together with matching guides. Both pins and guides have attached mounting plates by which they can be bolted, screwed or welded to the halves, making up a complete flask set.

basic pig iron: A special high phosphorous, low sulphur, low silicon pig iron made for open hearth process of steel making.

bath: The word bath refers to the molten metal on the hearth of a furnace during the melting process.

battens: Wooden bars or strips attached to patterns for rigidity and to prevent distortion during ramming of the mold (Fig. B-8). Thin flat patterns are often stiffened with battens for strength after the mold is made. The recess left in the mold by the battens is stopped off with a core or molding sand and sleeked smooth. *Stiffeners*, runner bars on bottom boards, follow boards, etc., are often called *battens*.

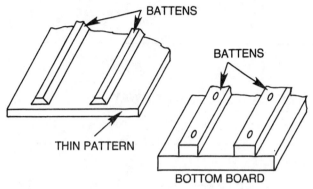

Fig. B-8. Battens.

bauxite: An ore of aluminum, moderately pure hydrated alumina $Al_2O_3 \cdot 2H_2O$. Theoretically 74 percent alumina. Bauxite has a very high melting point (1820°C.) and can be used directly as a refractory material.

bay: The longitudinal division of a foundry floor (Fig. B-9). The area between the bays is called the *gangway* or *gangways*. The various sizes of castings produced are usually divided into bays.

bayberry wax: A wax made from bayberries—an old-timer in the foundry. It is used as a liquid pattern parting by dissolving it in white gasoline. The pattern is coated and when the gasoline evaporates, a hard thin coating of bayberry wax is left on the pattern. It is also used to coat cast iron castings to prevent rusting and as a molder's shovel coating to prevent sand from adhering to the shovel.

beam and sling: The tackle arrangement used in conjunctions with a crane for turning over a cope or drag of a mold prior to assembly (Fig. B-10). Large flasks are fitted with *roll-over trunnions* over which the bails are placed to roll-over the flask section. The beam has notches in order to adjust the distance between bails for the various lengths of flasks.

The beam should be constructed so that the slings cannot be jarred off of the beam when in use. At any rate the notches in the beam should be deep enough. A hook on the ends of the beam adds to the safety factor.

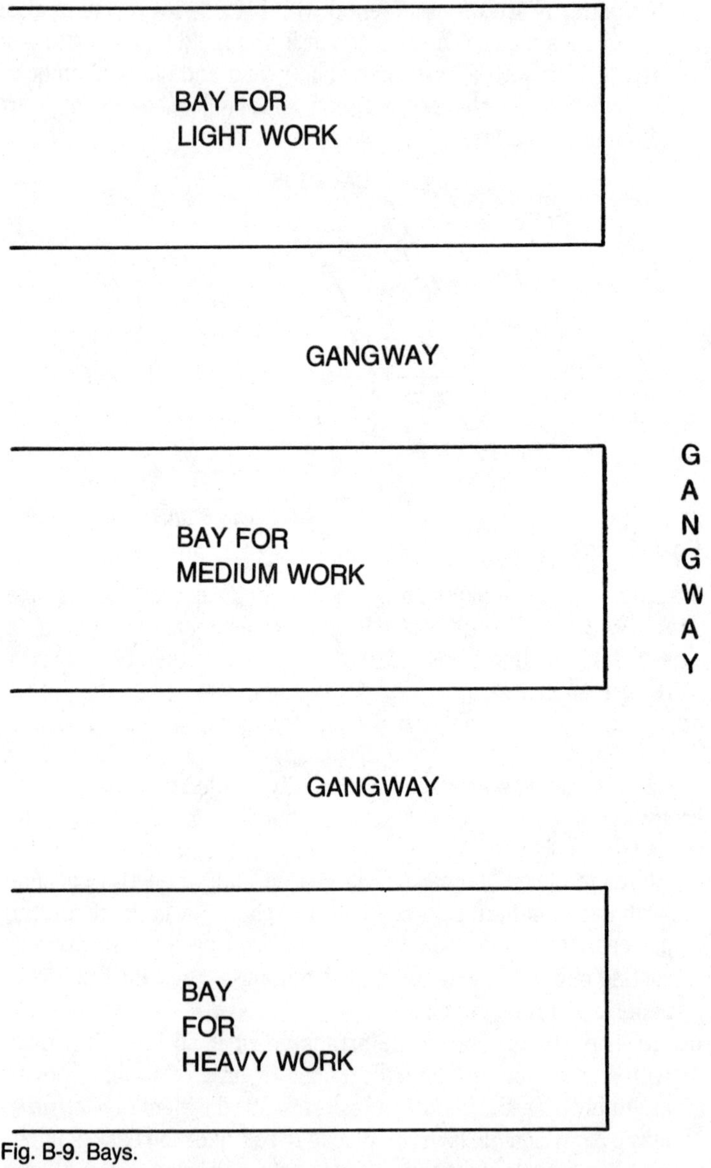

Fig. B-9. Bays.

bearing metals: Any metals used as bearings, ferrous and non-ferrous. There are hundreds of metals designated as bearing metals—from babbitts to oil-impregnated coppers.

bed: The initial charge of coke placed in a cupola extending above the tuyeres on which the metal charges are melted. The bed is

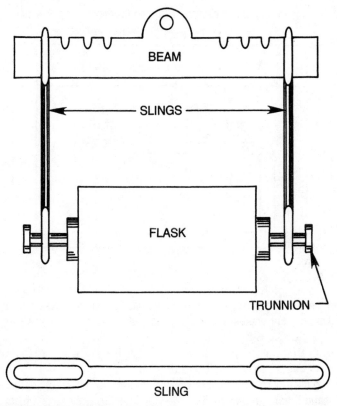

Fig. B-10. Beam and sling arrangement.

replaced and maintained at a fixed height by the coke between metal charges (continuous melting).

bedding in: With large patterns it is the practice to place the drag side of the pattern in the pit or drag flask on a bed of lightly rammed sand (Fig. B-11). Ram and tuck the sand under the pattern and around the sides working up to drag parting line. The pattern is usually lifted out prior to making up the cope so that the drag mold can be checked for soft spots and also for the parting. When satisfied that the drag is right, the pattern is returned and the cope made up in the usual manner. This system eliminates the necessity of rolling over the drag—and advantage with the production of large heavy castings. In some cases the pattern is supported during the bedding-in operation with an *overcastle*. The overcastle holds the pattern in the correct position during ramming and it is then removed to make up the cope.

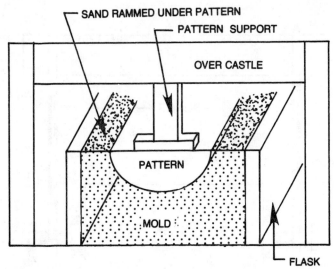

Fig. B-11. The process of bedding-in.

beeswax: Beeswax is used in wax fillet formulas and various investment waxes. It is a long chain ester, myricil palmitate and cerotic acid with a melting point of approximately 145°F.

bench lifter: A simple steel tool with a right-angled square foot on one end of a flat bent blade (Fig. B-12).

   This tool repairs sand molds and lifts out and any tramp or loose sand that might have fallen into a pocket. To remove dirt from a pocket that will not blow out with the bellows, you simply spit on the heel of the lifter and go down and pick it up, then wipe off the heel and go back and slick the spot down a bit.

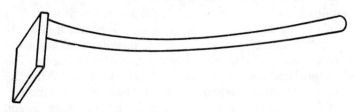

Fig. B-12. Bench lifter.

bench molder: A molder who produces small molds made on a molder's bench.

bench molds: Small hand-rammed molds made on the molder's bench that can be handled by one man. When they get too large, they are then made on the floor and called *floor molds*.

192

bench rammer: Made of oiled maple. You can buy one or turn one on a lathe and bandsaw the peen end with its wedge shape (Fig. B-13). The butt end is used to actually ram the sand tightly around the inside perimeter of the flask to prevent the cope or drag mold from falling out when either half is moved, rolled over a lifted.

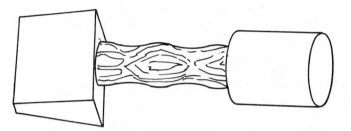

Fig. B-13. Bench rammer.

bentonite: A hydrous silicate of alumina, a colloidal clay derived from volcanic ash. It is employed as a binder in foundry sands and is added to naturally bonded sands to increase their green and dry strength. It is also used as an oil well mud and numerous other commercial uses. A typical green sand formula uses bentonite as the principal binder. By parts it is 14 parts silica or heap sand, six parts bank sand, one part air float sea coal and one part bentonite.

Bentonite is an unusual type of clay in that it swells 15 to 20 times its volume in water.

Two basic types are sold for foundry use—southern and western. The basic chemical difference between the two is that the western bentonite is higher in soda, alumina and chemical water (combined H^2O) and lower in potash, lime, magnesia and iron. Southern bentonite has less baked and hot strength. It collapses more readily, making it more suitable as a binder for light aluminum and bronze castings, where desirable. With steel and heavy castings, western bentonite will give the added hot strength and retard premature collapsability where needed.

It is a common practice to use the two blended together to produce the desired results for a particular type and weight of castings. The most common appears to be 50/50.

beryllium bronze: An alloy of copper and beryllium usually not more than 3 percent beryllium. Also called *beryllium copper*, it is used to produce strong mechanical parts, locomotive bearings, plastic molds, non-sparkling tools, springs, etc. Beryllium bronze is defined as an alloy containing over 2 percent of beryllium or beryllium plus other metals other than copper (over 2 percent).

No flux is necessary in melting beryllium bronze. A beryllium itself is a good deoxidizer. Some melters do use a dry charcoal cover when melting. The pouring range varies somewhat with the percent beryllium in the alloy being cast (Table B-1).

Beryllium bronze should be handled like aluminum bronze because it has a tendency to dross.

Per Cent Beryllium	Pouring Range
1.90 to 2.15	1900/2100° F.
2.50 to 2.75	1850/1900° F.
.45 to .60	2050/2200°F.

Table B-1. Beryllium Pouring Range.

bessemer ladle: A steel ladle lined with refractory material and having one or more stopper and nozzle systems for bottom pouring (Fig. B-14). The stopper is attached to the end of the stopper rod which is insulated with refractory sleeves. The stopper plug is seated in the nozzle when closed. The nozzle is opened to pour or teem by raising the stopper rod by a lever attached to the outside of the ladle and the top of the stopper rod.

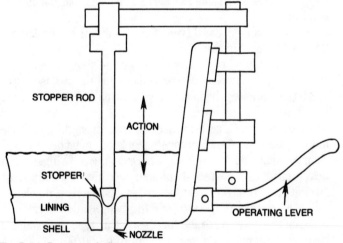

STOPPER ROD

ACTION

STOPPER

LINING

SHELL

NOZZLE

OPERATING LEVER

Fig. B-14. Bessemer ladle.

bessemer process: A steel-making process that blows air through or over the molten pig iron in a converter, oxidizing the silicon and carbon (Fig. B-15). The carbon is reduced to carbon monoxide and liberated as gas and the oxides of maganese and silicon form with the slag. The reactions which take place are exothermic and no additional heat is necessary in the refining process which is called the *blow*. It takes approximately 20 minutes to blow a heat. When the heat is finished it consists of some iron oxide, .03 percent silicon and manganese and approximately .06 percent carbon. The desired carbon content is adjusted by adding the proper amount of cupola iron and adjusting the composition with ferro alloys. The converter is mounted on trunnions so that it can be tipped for charging, blowing and pouring. The air is fed to the tuyeres through one of the hollow trunnions. In the side-blow converter the air is blown across the top face of the charge and the angle is adjusted by the operator. In the bottom blow converter the air is blown in through the charge from the bottom.

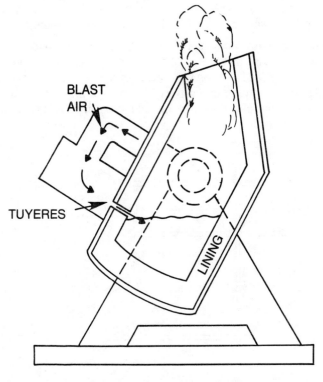

SIDE BLOW CONVERTER

Fig. B-15. Bessemer process.

binder: Any material used to hold a mold or core material together in the green or dry state such as cereal, pitch, resin, oil, sulphate, water glass, latex, etc. In an oil sand core, the oil is the binder. When baked, it cements or glues the sand grains together.

blacking or mold wash scab: A case where the blacking or wash on the mold or core, when heated, breaks away and lifts off of the surface like a leaf and is retained in or on the metal. The cause is a poor binder in the wash, an improperly dried wash, a poor wash formula or all of these possibilities.

blackstrap foundry molasses: A heavy, nonedible molasses which has been treated chemically to prevent it from souring or fermenting. It is used as a binder in core and mold washes, sprays, pastes, dough rolls and core daubing. Also used in skin-dried work by tempering the facing sand with a mixture of one part molasses to 10 parts water. It is a good spray at 10 to 1 for green sand cores which have been dried at 350° to 400°F. Sprayed while hot, it gives the core a nice firm skin.

blacking scab: A casting defect. It is caused by mold blacking or wash flaking off of the mold surface due to sand expansion and being retained in or on the surface of the casting. The cause is usually due to an excessively strong bond in the wash or not enough cushion material in the sand or wash. It is also related to

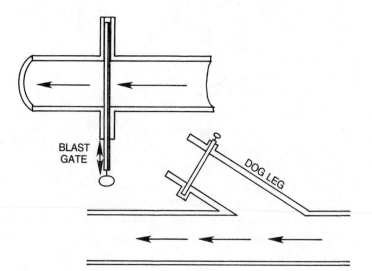

Fig. B-16. Blast gate.

backing holes which are irregular-shaped surface cavities containing carbonaceous matter.

blast: The wind or air supplied to a cupola, converter or furnace to support combustion.

blast cleaning: Cleaning (removing sand or oxide scale) by impinging action of sand, grit or shot projected by air, water or centrifugal force.

blast gate: A sliding metal gate used to adjust the flow of air to the cupola wind belt, and the air to burners on some types of nonferrous furnaces (Fig. B-16). There are two schools of thought as to where it should go. The first school says in the blast pipe proper and the second school says in a dog leg from the blast pipe to spill the air from the blower. Thus when closed, the air flows fully through the blast pipe and when open, less flows through the blast pipe. With the second arrangement the blower operates at a constant speed and load. With the first, when it is closed or partially closed there is a back pressure on the blower.

bleeder: A defect caused by shaking the casting out too soon when a portion of it is still liquid. The section runs out leaving a defect. In extreme cases the entire center section will run out on the shake-out man's feet.

blended sands: Use the suggested sand blends in Table B-2 as a guide in selecting your own basic blends. Start with a good high grade silica, washed, dried, screened and graded. Adjustments can be made to the base after sufficient tracking is done.

Table B-2. Suggested Blended Sands.

Substance	Fineness	Permeability
Heavy iron using green or dry sand	61 and 50	80 to 120
Medium iron using green sand.	70 & 45	50 to 70
Light squeezer iron, green sand	110 & 80	20 to 30
Stove plate iron, green sand	200 & 160	9 to 17
Heavy green steel, green sand	60 & 35	140 to 290
Heavy steel, dry sand	55 & 40	90 to 250
Light squeezer malleable iron	130 & 95	20 to 40
Heavy malleable iron	80 & 70	40 to 70
Copper and monel	150 & 120	30 to 60
Aluminum	250 & 150	6 to 15
General brass	150 & 120	12 to 20

blind riser: A blind riser is any feed riser which does not extend through the cope to the atmosphere.

blind risers (foam): Blind risers are formed in a sand mold by using styrofoam pre-formed spheres and pear shapes which are stuck down on a locating pin on the pattern at the riser sight (Fig. B-17). The spheres and pears have a pre-formed recess which fits down on the locating pin. In operation the desired size of the sphere or pear is placed on the locater pin and the mold is rammed in the usual manner. When the cope is drawn, the styrofoam blind riser remains in the sand body. The metal poured into the mold vaporizes the styrofoam, forming the riser to feed the casting. It is a common practice to vent the riser former with a ⅛ to 3/16 vent wire through the cope. The pin on

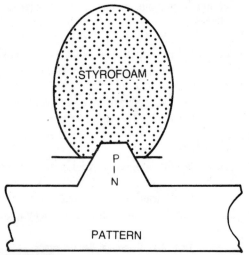

Fig. B-17. Foam blind riser.

the pattern should be taller than the depth of the retaining hole in the styrofoam riser former so that the styrofoam shape winds up above the pattern slightly. The pin length will vary depending upon whether the mold is hand rammed or rammed on a squezzer and how close the styrofoam form is to the top of the cope.

Spheres can be purchased in diameters from 2½ inches to 6 inches, or in pear shapes from 2½ inches to 6 inches. They may be purchased from Frostee Foam Co., Box L, Antioch, IL 60002. Or, you can fabricate your own.

blister: A shallow blow covered over with a thin film of metal.

bloating: The swelling of any refractory material due to excessive heat or carbon monoxide in the furnace.

blowers: Blowers come in all sizes and shapes (Figs. B-18 and B-19). They are rated in cubic feet per minute at various delivery

Fig. B-18. A positive pressure blower.

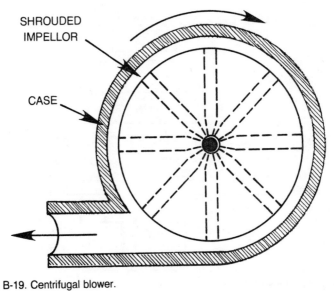

Fig. B-19. Centrifugal blower.

pressures. The most common type of blower is the *centrifugal fan type*—a cast or fabricated case which turns a paddle type of impellor. The design of these blowers varies considerably from manufacturer to manufacturer and also with the intended use for a particular blower. Blowers are often called *fans* in the foundry. In some blowers the impellor itself is *shrouded* on both sides with the idea that this increases its efficiency, both in pressure and volume. This question is up for grabs.

The shrouds are riveted or screwed to the edges of the blades. Some blowers are compound. One stage delivers air to another stage, upping the output.

The other common type of foundry blower is the *positive pressure blower*. This type of blower has two lobe-shaped impellors which are geared to each other to maintain an exact relation between them at all times. They do not touch each other or the case, but are fitted so closely that it is impossible for the air once taken in to pass back through the intake.

The positive pressure blower is superior to the centrifugal fan type for supplying air to a cupola as the pressure remains constant regardless of conditions in the cupola.

blowing of cores: Cores are often produced on machines called *core blowers* which vary greatly in size, capacity and mechanization (Fig. B-20). However, regardless of the blower design and size, the principal is the same. The sand mix is blown into the core

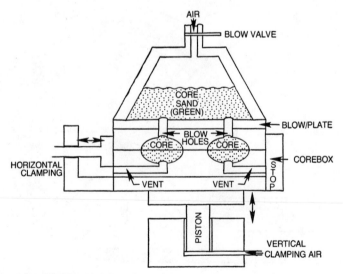

Fig. B-20. Blowing of cores.

box by introducing air at high pressure on the sand in a reservoir, thus blowing the sand into the core box through blow holes in the top of the core box and through the blow plate on the sand reservoir. The sand travelling at a high velocity is abruptly stopped by the cavity walls of the box, thus ramming itself. The air continues through the sand and out the vents in the core box.

The machine is provided with the mechanical arrangements for moving the reservoir to fill it and clamping the box vertically and horizontally as needed during the blow cycle. The controls are tied together in such a way that the blow cannot be made until the box is in position and properly clamped. The reservoir is exhausted prior to releasing the box.

blows: Round to elongated holes caused by the generation or accumulation of entrapped gas or air. The usual cause of blow holes is:

- sand rammed too hard (decreasing the permeability)
- permeability too low for the job
- sand and core too wet (excessive moisture)
- insufficient or closed-off core vent
- green core not properly dried
- incompletely dried core or mold wash
- insufficient mold venting
- insufficient hydrostatic pressure (cope too short)
- cope bars too close to mold cavity
- wet gagger or soldier too close to mold cavity
- poor grain distribution

Any combination of hot and cold materials would lead to condensation as a hot core is set in a mold or vice versa. A cold chill or hot chill will cause blows, as well as wet or rusty chaplets.

blowy metal: When a metal is gassed in melting to the point that it kicks and churns in the furnace or ladle, it is referred to as blowy metal and should not be poured unless it can be saved by degassing or deoxidizing. With most metals simply re-melting again under the correct conditions will correct the problem. Metal poured into a wet or incompletely dried ladle will kick and blow. Trying to dry a ladle with hot metal just won't work.

bobble: The casting will resemble a cold shut or look like it was poured short. The problem was a slacked or interrupted pour. When pouring a casting the metal must be poured at a constant choked velocity. If you slack off or reduce the velocity, you can cause this defect. If you must completely interrupted pour, start—stop—and start, only for a second. A great percentage of lost castings is caused by pouring improperly.

bolted cement: A super fine *Portland cement* used for *print back* or *return facing.* In green sand casting where an exceptionally fine surface detail is required as in casting art work, plaques and grave markers the practice is to draw the pattern, dust the mold cavity lightly with bolted cement and immediately return the pattern to the cavity and tap down. Then you redraw the pattern. This process sets the cement smoothly in the cavity. The mold is closed and poured in the usual manner. If done properly, you can produce some very fine work.

bolted charcoal: Finely ground charcoal used in a *dust bag* as a shake on mold facing and in print back work. Sometimes it is combined with graphite. It can be used as a parting dust. *Granulated charcoal* is used as a cover on the surface of molten metal, especially brass, to prevent metal oxidation from the products of combustion.

bolted cores: The practice of bolting core together where it is felt that pasting could fail due to the pressure involved in pouring a particular (design) casting (Fig. B-21). The cores are drilled and countersunk or provisions are made in the core box to form the bolt holes. After bolting, the counter sink recess must be filled and mudded. Separate cores can be made and baked and simply glued in place. The joint is then mudded.

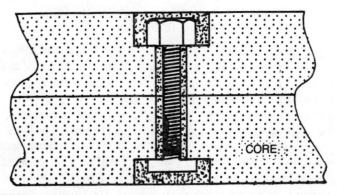

Fig. B-21. Bolted cores.

bolting pads: A pad or boss type projection on a casting, which if not properly designed or chilled, will produce a hot spot and localized skrinkage at that point due to the abrupt change of metal thickness (Fig. B-22). *Hot spots* are also caused by raised pattern letters and incorrect rib design. If the bolting pads are large enough they can be cored out to prevent localized shrinkage.

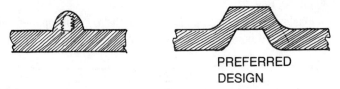

PREFERRED
DESIGN

Fig. B-22. Bolting pads.

bonding action of clay in molding sand: Although most clay is sticky when moist, the sand grains of the mold are held in place by forming clay wedges holding the sand grains in place, as done for a stone wall—the old wedge and block theory (Fig. B-23).

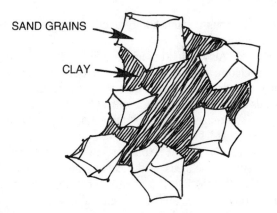

SAND GRAINS

CLAY

Fig. B-23. The bonding action of clay in molding sand.

bonding clays: A large family of clays used in bonding molding sands composed of one or more of the minerals, kaolinite, illite and montomorillonite (Bentonite).

Clays came about from the weathering of various types of volcanic ash. Clays are composed of minute crystalline flakes of constituent minerals. Generally the smaller the size of the particles of the clay minerals, the greater the bonding power.

booking: Booking refers to green and dry sand cores (Fig. B-24). Using a full-core box, ramming each half box with sand and striking both off level with the box and closing them together like a book brings the mating surfaces at the parting line together. Then remove one half box and place a *drier* in place of the half box that was removed. Roll the remaining half box and drier over and remove the box leaving the core resting in the drier for baking. In the case of a green sand core on a half arbor, remove the core from the box.

Flasks with *roll-off hinges*, where the cope is opened and closed like a book, are referred to as *booked molds*.

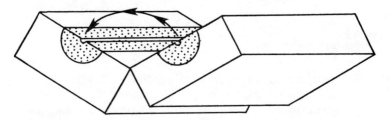

Fig. B-24. Booking process.

borax: Hydrous sodium borate used in the foundry as a flux in melting metals.

boric acid: Also called baracic acid B_2O_3. $3H_2O$, made by adding hydrochloric acid or sulphuric acid to borax and crystallizing. Its main foundry use is as a protective agent in magnesium molding sands usually in combination with sulphur or sulphur and fluoride salts. The reason is to inhibit the action of water vapor upon the hot metal (Magnesium).

The amount used is 3 to 6 percent by weight depending upon the section thickness of the casting and the sand. The more massive the casting, the greater the amount needed. Also the injection of a small amount of sulphur dioxide gas into the mold just before pouring helps to prevent oxidation of the surface of the casting.

bott: A blunt cone of clay used to stop off the flow of iron from the tap hole of a cupola or air furnace. There are many mixes usually formulated by the melter. One such mix is: five parts molding sand, five parts clay and one part wheat flour.

Some tenders, after forming the bott and placing it on the *bott rod*, sprinkle a coating of coal dust or graphite on the bott with a shake bag. This seems to help release the bott when tapping, providing there is a parting between the tap hole and bott.

bott stick: More often called a *bott rod*. It is a rod ¾ inch in diameter 8 feet long with a 2 inch diameter plate welded to the business end and a loop handle on the other (Fig. B-25). It is used to place the bott firmly into the tap hole of a cupola or air furnace to shut off the flow of metal. The bott is stuck to the flat plate by wetting it (the plate) and pressed firmly into the tap hole, coming in at a slight angle above the metal stream. When in place, the bar is twisted and slid downward to release the bott from the rod.

The English Bott rod is domed on the end. The theory is that a large surface area will hold the bott and there will be less likelihood of losing the bott (of it falling off).

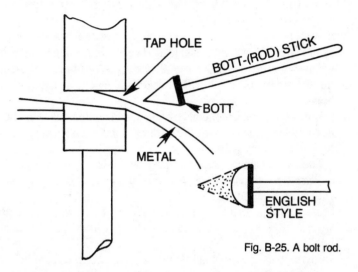

Fig. B-25. A bolt rod.

bottom board: The board used to support the sand mold until the mold is poured. It is constructed like the molding board, but need not be smooth—only level and stiff. Bottom boards from 10 inches × 16 inches to 18 inches × 30 inches should be made of 1 inch thick lumber with two cleats made of 2 × 3 inch stock. Bottom boards 48 × 30 inches use 1½-inch thick lumber with three cleats made of 3 × 3 inch stock. Bottom boards 50 to 80 inches use 2-inch thick lumber with four to five cleats made of 3 × 4 inch stock.

bottom pouring: Filling the mold cavity from the bottom by simple displacement by means of gates entering the bottom of the mold from the runner.

bottom sand: The molding sand rammed against the bottom doors of a cupola to form the sloping hearth. Also called the *crucible bottom*.

Some brass melters ram a molding sand bottom in a brass furnace around the pedestal block. This makes it easy to clean out the bottom for maintenance or to remove a spill in the bottom.

bouyouzos: A hydrometer calibrated in grams per liter. The hydrometer shows the number of grams still in suspension at the time it is read. Used to determine the size distribution of small particles.

brads: Small, usually not headed, wire nails used in pattern and model work. When headed they are then called *flathead wire nails*. Brads run from ¼ inch long, 22 gauge to 3 inches long, 12 gauge.

brasses: Red brass, leaded red brass, semi-red brass, leaded semi-red brass, yellow brass, leaded yellow brass, high strength yellow brass, leaded high strength brass, silicon brass, tin brass, tin nickel brass, nickel brass (nickel silver) and leaded nickel brass.

brazing: When you join metals by fusion of a nonferrous alloy that has a melting point above 800°F, but at a lower fusion point than the metals being joined.

The three basic types of brazing methods are *torch brazing, furnace brazing* and *dip* or *flux brazing*. Torch brazing is done with a torch and rod (as filler metal). For furnace brazing and dip brazing, the parts are assembled and the filler metal applied as wire, washers, clips and bands. Or they may be integrally bonded.

Silver soldering is actually silver brazing. When you get below 800°F. melting point (with your filler metal),you are soft soldering.

breast: The refractory portion of a cupola or air furnace in which the *tap hole* is located (Fig. B-26). It is located in the front of the cupola and on each side of an air furnace usually equipped with a *pouring spout* which extends outward from the furnace to direct the stream of metal from the furnace into the receiving ladle.

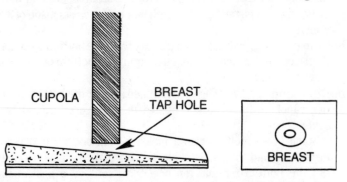

Fig. B-26. The breast of the refractory portion of a cupola or air furnace.

breeze: Coke and coal screenings used to vent large cores and molds (Fig. B-27). Widely used in large sweep work and large skeleton (pattern) work.

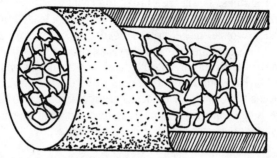

Fig. B-27. A large core filled with breeze.

bridge chaplet: Bridge chaplets are made of 20 gauge cold rolled steel and hot dip tinned or copper plated (Fig. B-28). An economical chaplet for use in castings where the conditions under which the chaplets are used are not severe.

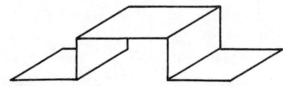

Fig. B-28. Bridge chaplet.

brinnell hardness tester: A device (press) which presses a 10MM diameter steel ball against a metal's ample or casting at a known or given load—500 Kg for nonferrous, 3000 for ferrous. The diameter of the indention made on the specimen in question is measured and the hardness number is taken from a table computing the hardness.

briquets: Compact or cylinerical-shaped blocks formed of finely divided materials. They are under pressure with various binders in order to increase their density and reduce their loss due to oxidation and draft in melting or treating a bath. Includes borings, chips, ferro alloys, silicon carbide, charcoal dust, coal dust, etc.

broken casting: Often caused by improper design or improper filleting along with improper handling anywhere along the line. copper base castings, red brass, yellow brass, etc., are what are known as *hot short*—breaks easily when hot. Thus, if the casting is shaken out of the mold before it has cooled sufficiently, it can get broken very easily. A mold or core which has a too high *hot strength* will not give or collapse to give the casting room to move as it shrinks. It will then break the casting.

bronzes: Tin bronze, leaded Tin Bronze, high leaded tin bronze, leaded bronze, nickel bronze, leaded nickel bronze, aluminum bronze, silicon bronze and beryllium bronze.

Brunelli strainer: Invented by an Italian foundry engineer (Fig. B-29). His claim is that his pouring system does away with the necessity of risers or feeders on large castings. He also claims the metal entering the mold is oxidized in proportion to the volume of air with which it comes in contact.

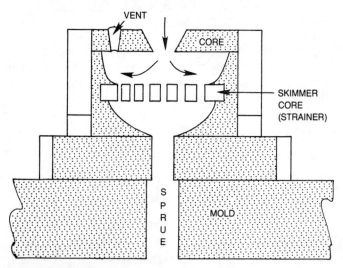

Fig. B-29. Brunelli strainer.

buckle: An expansion scab, an indentation on the face of the casting caused by the sand being rammed too hard or not containing sufficient combustible material such as wood flour or sea coal (Fig. B-30). what happens is that the heat from the metal causes the sand to expand and it cannot do so without buckling. It is a very common defect in full-mold investment casting when insufficient gating is used and the mold fills too slow, subjecting the mold walls to the heat for too long a period of time causing excessive expansion and spalling.

bulb (paste): A short-snouted rubber bulb used to apply core paste to the joints of cores for pasting them together (Fig. B-31). Used

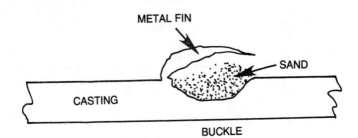

Fig. B-30. A buckle.

to run a paste line on the drag of small molds to help prevent a run-out at the joint and over the print on cores to stop metal from running over the ends and stopping off the core vent. Also used to put paste on the handle of the finishing trowel in the back pocket of the molder. When he reaches for his trowel, he will get a good grip.

bulb sponge: The bulb sponge consists of a rubber bulb with a hollow brass stem terminating in a soft brush (Fig. B-32). The stem is pulled out and the bulb filled about three-fourths full of water. The stem is then replaced. The bulb sponge is used to swab around the pattern prior to drawing it in the way the flax swab is used for floor molding. The bulb is gently squeezed to keep the brush wet while swabbing.

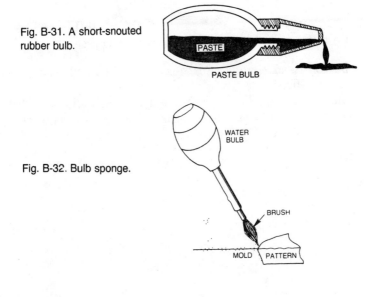

Fig. B-31. A short-snouted rubber bulb.

Fig. B-32. Bulb sponge.

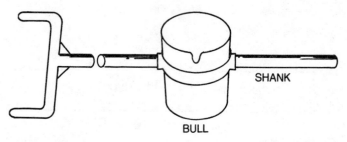

Fig. B-33. Bull ladle.

bull ladle: A two-man ladle for carrying and pouring metal. It has a shank with two handles. When ladles get much over 200 pounds capacity, they become *crane* or *trolley ladles*.

bungs: A section of a removable arch roof. In large air furnaces and kilns the roof is very often made by using cast iron frames which carry a section of the arched roof. The frames have pat eyes for lifting them off with a crane for maintenance, charging, etc.

burned sand: Molding sand in which the clay portion has been destroyed by the heat of the metal.

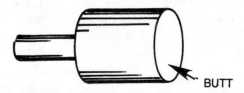

Fig. B-34. Butt ramming is done with a butt-shaped rammer

button head chaplet: Round-headed chaplet with a forged or upset head.

butt ramming: Ramming the cope or drag with a butt-shape rammer, either hand or pneumatic. The butt rammer is the flat circular end of the rammer.

C

cabbage head: A puffed riser head caused by badly gassed metal (Fig C-1). In copper casting the molder usually pours what is known as a *text cock*—a sprue cut into a box rammed with sand. If a cabbage head forms, the metal is deoxidized again and a second cock is poured. This process is repeated until the cock pulls (shrinks), then and only then is the mold poured. Very good insurance when pouring a large mold. A good practice with any metal which is subjected to gassing.

calcium boride: An alloy of calcium and boron Ca B$_6$ used as a deoxidizer of nonferrous metals. Often the prime ingredient in a commercial deoxidizer.

calcium carbide: Produced by fusing lime and coke in the electric furnace. When calcium carbide is put into water (or gets wet) it produces acetylene gas and a residue of slaked lime. Widely used as a welding and cutting gas when combined with an oxygen oxy/acetylene torch.

calcium manganese-silicon: An alloy used as a scavenger for oxides, gasses and nonmettalic impurities in steel.

calcium molybdate: A crushed product of lime, molybdenum, iron and silica used to add molybdenum to iron and steel heats.

calcium silicon: An alloy of calcium, silicon and iron used as a deoxidizer and degassifier for steel. Also known as *calcium silicide*.

calcium sulfate: The principal binder used in full investment molds for casting nonferrous metals. Gypsum, when calcined, is

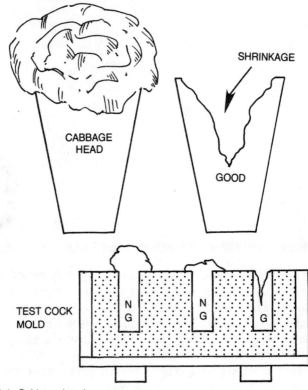

Fig. C-1. Cabbage head.

called plaster of paris. Typical formula for plaster bonded investment is two parts plaster, three parts silicon flour, olivine flour or crushed red bricks.

camber: As a defect it is a warped casting. It is caused by any number of problems: incorrect number and size of ribs; a design which contributes to differential stress; flask bars too close to the pattern; sand too strong in green, dry or hot strength; or insufficient combustibles in the sand (wood flour etc.).

camber (pattern): A pattern is intentionally bowed, sprung or faked to prevent a camber defect (warp) (Fig. C-2). The pattern is cambered in the opposite direction of the anticipated warp. If done correctly the casting will come out true.

In patterns for cast iron oil pans and sumps for tractor engines, the pattern is often cambered in order to produce a straight flat surface where the pan mates up with the engine.

compound camber: A camber which is both bent and twisted.

captive foundry: A foundry that is part of a manufacturing company or operation which consumes its entire output and produces no castings for outside customers.

carbide: A compound of carbon with one or more metallic elements such as silicon carbide, aluminum carbide etc.

carbon dioxide: CO_2 A colorless, odorless gas. Also called *Carbonic anhydride* when solid dry ice. It is used to cure (set the binder) when making molds and cores using silicate of soda (water glass) as the binder by passing the gas through the sand and silicate of soda mix.

carbon equivalent: The relationship of the total carbon, silicon and phosphorous content in a gray iron expressed by the formula:
$$C.E. = T.C. + \frac{Si\ percent + P\ percent}{3}$$
T.C. = Total carbon − Si = Silicon − P = Phosphorous.

carbon (in gray iron): Carbon in gray iron is in two forms, combined (in solution) and free (graphite flakes). The ratio of combined to free carbon is controlled by the charge make up and silicon content. Silicon throws the combined carbon out into the iron in the form of free carbon. The strength properties are improved by the reduction of free carbon flakes and the uniform

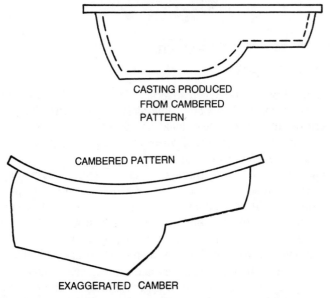

CASTING PRODUCED
FROM CAMBERED
PATTERN

CAMBERED PATTERN

EXAGGERATED CAMBER

Fig. C-2. Chamber patterns are intentionally bowed to prevent defects.

disturbing on. The smaller the flake size the better. The silicon lowers the solvent power of iron for carbon.

carbon steel: Steel owes its properties chiefly to its carbon contents. It is without substantial amounts of other alloying elements. Also called *ordinary steel*, *straight carbon steel* and *plain carbon steel*.

card pattern: A number of small patterns fastened together by gates or to themselves (Fig. C-3). Bearing caps and clutch dog patterns are often carded and cast as a unit and sawed apart to produce the upper and lower member.

Bearing cap carded patterns have the advantage that the bore is machined before the castings are parted. In this way the radius of the upper and lower member is exactly the same as being machined at the same time on the same center.

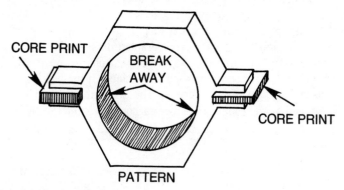

Fig. C-3. Card pattern.

card of patterns: When several different loosely gated patterns are assembled as a unit to be molded in the same flask.

car oven: A large core or mold drying oven with tracks on which a large car (containing the cores or molds) can be pushed (rolled) into the oven for drying.

cast gate: The gate through which the metal enters the mold cavity. Also a gate cast to a metal pattern as part of the pattern.

cast steel: Any object made by pouring molten steel into molds.

casting design: The reference here is that an item to be cast must be designed with the selected casting method in mind. This requires a good working knowledge of the casting methods and practices. If not, the foundry and pattern maker should be consulted which will result in a great saving to the designer. It involves many, many factors such as metal selection, pattern equipment, casting method, finishing, etc. The casting cost can often be greatly increased by overlooking some simple element.

casting ladle: Any device used to transfer molten metal into a mold, as opposed to a *transfer ladle* which is used to feed the casting ladles.

casting strains: Internal strains resulting from the cooling of a casting plus residual stresses. Strains in castings can be removed by various heat treating and stress relieving techniques. The design along with the particular metal poured plays a great part in strains and stresses in a casting.

cast structure: The structure of a cast alloy consisting of cored dendrites and in some alloys a network of other constituents. The structure on a microscopic scale. A micrograph structural study.

cast weldments: The reduction of a complicated casting into two or more simple castings and the welding together to produce the desired part.

caustic dip: A solution of sodium hydroxide (lye) and water used to clean the surface of a metal. Bright dip for aluminum castings. Also used as an etch with aluminum alloys to reveal the microstructure.

cavitationerosion: Erosion of a material due to the formation and collapsing of cavities in a liquid at the solid-liquid interface. The principal of ultra sonic cleaning in a liquid.

cement bonded sands: Molds and cores produced by bonding silica sand with hi early cement and air drying for a minimum of 72 hours. Also known as *72 hour sand*. The first cement sand mix was patented by John Smith of Verona, Pennsylvania over 70 years ago. It consisted of 10 parts silica sand and one part portland cement moistened to 6 percent moisture with molasses water. It was used to cast both ferrous and nonferrous metals. Still used to some extent today, but largely replaced by Furan no-bake mixes.

cementite: The structure with the compound Fe_3C which forms when the Austenite changes on cooling. The combined carbon structure of cast iron is known as *austenite*.

center line: A well defined gauge line placed on the work or layout to serve as a basis from which dimensions are to be measured.

centrifugal casting: The production of castings by centrifugal force (true centrifugal) where the molten metal is poured into a rotating mold at the axis of rotation and is conveyed by centrifugal force out to the peripheral extremeties leaving you with a hollow casting with the impurities in the center or bore (Fig. C-4). Cast iron and steel pipe are produced by this method as are bronze bushing stocks.

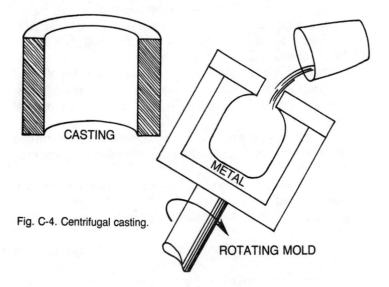

Fig. C-4. Centrifugal casting.

centrifuging: The process of casting whereby the metal is driven into mold varities which are located off the center of rotation (Fig. C-5). These can be single or multiple cavities. The mold material can be metal, rubber, green sand, investment, carbon, etc. More often than not, it is mistakenly called centrigual casting.

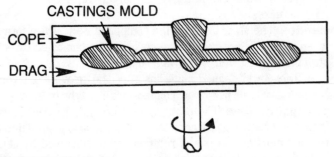

Fig. C-5. Centrifuging.

center line shrinkage: A defect which can occur in any metal (Fig. C-6). It is most often found in steel in thin sections although but not relegated solely to thin sections. A shrinkage cavity in the center of the casing wall due to the stopping of feed metal.

This shrinkage can be eliminated by tapering the casting to promote directional solidification toward the feed metal or padding. It can be removed by machining or grinding.

The author has seen cases of railroad trucks with centerline shrinkage to the point that they were hollow as if they were cored. They are not scrapped as it is a fact that a casting with centerline shrinkage is serviceable if the load is flexual.

The author cast a very large plaque 6 feet × 4 feet which was perfect in all respects except for it is completely hollow centerline shrinkage.

That was some 25 or 26 years ago and it's still hanging on the building. The metal was 88 percent Cu, 10 percent Sn and 2 percent Zn.

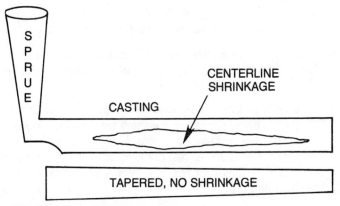

Fig. C-6. Centerline shrinkage.

cereal: A bonding material and a cushion material. The most widely used in the foundry is corn flour produced from wet milling of corn starch. The effectiveness of corn flour depends entirely on how it is used. In synthetic sands it works well from ½ percent to 2½ percent by weight. It is widely used in steel sands. The author believes in using corn flour in nonferrous sands to impart resilience. Corn flour increases dry compression strength, baked permeability, toughness, mullingtime, moisture pick up, mold hardness and deformation. In turn it decreases the green permeability, sand expansion, hot compressive strength and flowability. Cereals are replaced with wood flour to some extent, however, each has its distinct use. Corn flour is very beneficial when you have weak sands and no other additive can eliminate or reduce scabs, rattails, buckles and similar defects quite as well as 1 per cent to 1½ per cent corn flour added to the facing. It also keeps spalding to a minimum.

chamfer: To bevel a sharp edge (Fig C-7).

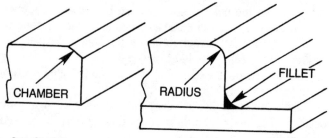

Fig. C-7 Chamfer

chamotte: A refractory produced from high alumina clays which have been calcined to a temperature above their softening point.

chaplet clips: Clips used to attach two or more chaplets together when the proper size is not available in your chaplet stock (Fig C-8).

chaplets: A large family of metal devices, supports or spacers used to support cores in molds long enough for the metal to solidify without moving or breaking a core during pouring (Fig. C-9.) They burn in and become part of the parent metal. Chaplets are not used when the pattern has sufficient core prints which serve the same purpose.

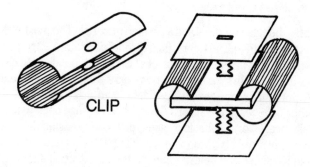

Fig C-8. Chaplet clips.

218

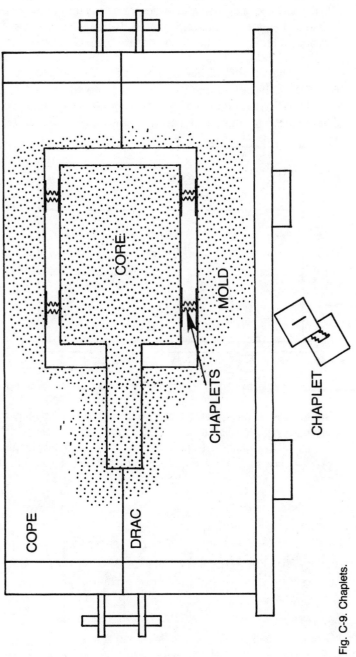

Fig. C-9. Chaplets.

219

charcoal: A high amorphous carbon fuel produced by heating wood in the absence of oxygen used widely to melt in the cupola at one time and as a cover material on meal baths, in core and mold washes, etc. At one time all iron was melted using charcoal as a fuel.

charge: The metal placed in a cupola or furnace for melting.

cheek: A section of flask between the cope and the drag sections of a flask to decrease the difficulty of molding an unusual shape or to fill the need for more than one parting line (Fig. C-10).

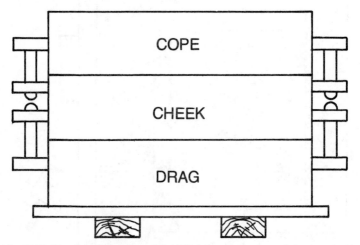

Fig. C-10. Cheek is the section of flask between the cope and drag section.

chill: Metallic inserts (usually cast iron) located in the mold to promote rapid cooling at a desired point to make the casting harder at this point or to promote directional solidfication and prevent shrinkage at a hot spot which cannot be fed from a riser (Fig. C-11). They can be internal as well as external.

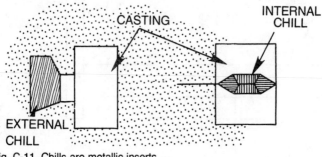

Fig. C-11. Chills are metallic inserts.

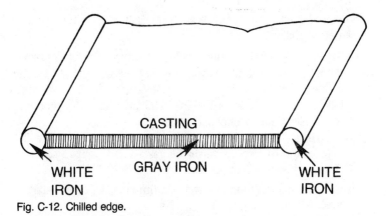

Fig. C-12. Chilled edge.

chilled edges: A condition (defect) in thin cast iron where the edges of the casting are hard (white iron). (Fig. C-12). The major cause is that the carbon equivalent is too low for the section being cast. If the CE is too high for the section it will *kish* (excessive free graphite at the edges and thin sections).

chill nails: Large-headed nails used as internal and external chills (Fig. C-13). Horseshoe nails are widely used due to the head mass. The nails are pushed into the sand at the point that is

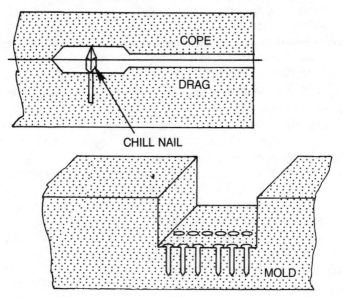

Fig. C-13. Chill nails.

desired to chill. Fanner chills are a specially manufactured nail for internal chilling.

chill wash: A wash coating applied to external chills to prevent them from kicking and blowing causing a defect at their point of contact with the casting. These wash coatings include washes of lube oil and graphite, lard oil; red lead and water (dry thoroughly); shellac; and alcohol and resin.

chloration: The process of passing chlorine gas through molten magnesium to remove dissolved gasses and trapped oxides (a degassing and refining process).

choking: Keeping the pouring basin brimming full from start to finish when pouring a casting.

Christmas tree: A single sprue containing many castings formed by a stacked mold.

chrome pickle: A pickling solution for dipping magnesium castings which produce a protective film on the casting and provide a base for painting. It consists of a solution of nitric acid and sodium dichromate.

chromium: An elementary metal Cr used as an alloy in stainless steels, heat resistant steels and high strength steels for plating, hard facing in on ferrous alloys, cast iron, etc.

churning: Also called *feeding* or *pumping*. A molder churns the liquid metal in a riser with a wrought iron bar to keep it open and assist in feeding the casting.

cinder mill: A revolving tumbler used to wash, cool and retrieve good coke and iron from the cupola drop. When the day's heat is over, the bottom doors on the cupola are dropped and the coke and iron remaining drop out. It is quenched with water and sent to the cinder mill.

clamp off: A defect caused by the displacement of sand in a mold due to clamping or improper handling of weights. It results in an indentation in the casting surface.

clay tile gates: The practice with large iron and steel castings to use pre-formed gate tiles of ceramic materials to prevent the washing and eroding of the mold and its gating system during pouring. It is basically a plumbing system of ceramic piping, elbow , tees and splash tiles. These tiles can be purchased in a wide vareity of sizes just like plumbing ware.

cleaning castings: Removing the sand and oxide scale and shaking out the cores by hand, sand blasting, shot blasting, chemical pickling etc.

This often represents a good chunk of the cost of produce the casting.

cleavage plane: The junction where two independent planes of crystal growth meet and interfere with each other (Fig. C-14). This condition produces a weak point at the junction. Its primary cause is a sharp edge or junction which is not filleted.

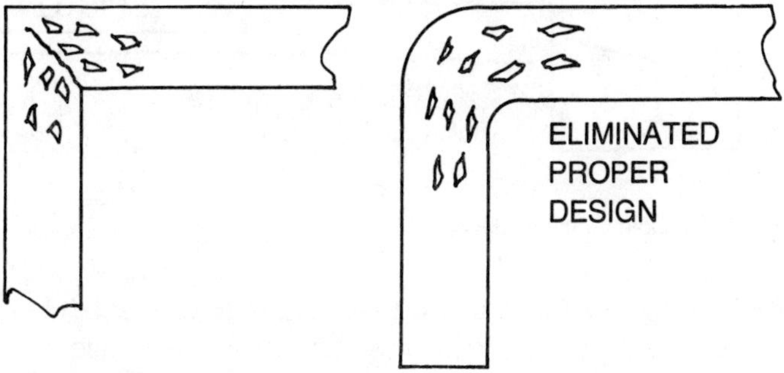

Fig. C-14. Cleavage plane.

cohesion: The force by which various particles are held together. Factors being there are hot or cold workings of the material and the molecular arrangement due to heat treatment of chemical action.

coke: The residue or product left which is mainly fixed carbon and ash produced by coking—heating bituminous coal in the absence of air to 1200 to 1400°C and expelling the volatile matter.

A good grade coal will produce from 1 ton of coal, .7 tons of coke, 11,5000 cubic feet of gas, 12 gallons of tar, 27 pounds ammonium sulphate, 50 gallons of benzol and .9 gallon of Tolool and Naphtha.

Foundry coke should have an ignition point of 1000°F sulphur, maximum .7 percent, and be strong enough to carry the charges in the cupola. Coke is widely used as the fuel in the blast furnace.

cold chamber die casting: In the cold chamber machine the metal is ladled into the shot chamber which is not emerged in the molten bath as the hot chamber machines (Fig. C-15).

cold shortness: Metals that are brittle at room or low temperatures are called cold short metals.

cold shot: Where two streams of metal in a mold coming together fail to weld together. This defect is usually caused by a metal too cold poured too slowly or a gating system improperly designed so that the mold cannot be filled fast enough.

223

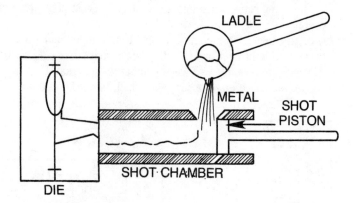

Fig. C-15. Cold chamber die casting.

collapsibility: A measurement of the ease or resistance of a sand mixture or core to break down under castings conditions.

A core in a casting must collapse as the casting cools and shrinks to prevent hot tearing and also the removal of the core. The degree of collapsibility, a factor of hot strength, varies depending on the pouring temperature, casting design, metal and the size of the casting.

The mold and core must collapse at the right time. A core in an aluminum casting must collapse at a fairly low temperature and much faster than one for a steel casting. It is controlled by the type, kind and amount of binder used.

colloids: Materials less than .0002 inches in size, gelatinous, highly absorbent and sticky when wet.

combination core box: A core box that may be altered with a plug or stop off to produce a core of another shape.

Such a box is one built by which a left-and right-hand core can be made from the same box.

combined carbon: The carbon which is chemically combined in iron and steel.

combined water: The water which is chemically combined in a mineral matter and can be driven off only by temperatures above the boiling point of water. The removal of combined water is called *calcining.*

combustibility: The ability of a substance to burn or combust. Combustion is the chemical change as the result of combining the combustible constituents of a matter with oxygen producing heat energy.

compressive strength: The maximum stress that a material can stand without a pre-defined amount of deformation when subjected to compression.

compound skimmer: A skimmer and pouring head arrangement combined (Fig. C-16).

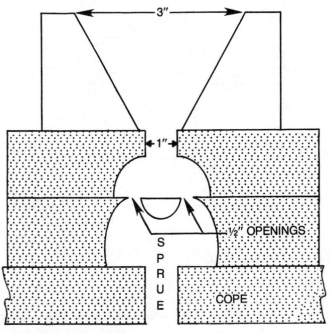

Fig. C-16. Compound skimmer.

condensation: Moisture formed from placing a hot core in a cold mold or a cold chill in a hot mold and causing a blow defect.

contraction: The reduction in size of material as it cools from an elevated heat to room temperature. The contraction of most bronze alloys from molten to room temperature is 3/16 of an inch per foot.

cope: The upper or topmost section of a flask, mold or pattern.

cope and drag mounts: In this case you have two separate pattern mounts, one fitted with female guides for the drag and one fitted with pins for the cope. These must match up with the flask used (Fig. C-17).

The cope half of the pattern is attached to the cope mount and the drag pattern is attached to the drag mount. The cope and drag molds are produced separately and put together for pouring. The usual practice is for one molder to make copes and another to make the drags. Cope and drag mounts are quite common when making large and very large castings where a matchplate would be out of the question due to its bulk and weight. Cope and drag mounts are sometimes called *tubs*.

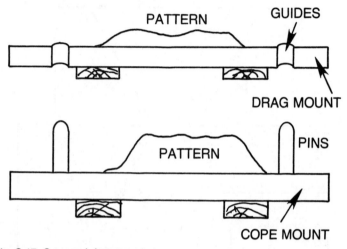

Fig. C-17 Cope and drag mounts.

coping out: Extending the sand of the cope downward into the drag by cutting away a portion of the sand in the drag prior to ramming the cope to facilitate the making of the mold and removal of the pattern by establishing the correct mold parting eliminate back draft (Fig. C-18).

copper: Yellowish red in color, tough, ductile and malleable metal. Found native in a large number of ores. Melting point, 1083°C; boiling point, 2310°C; specific gravity 8.91; and weight, .321 pounds per cubic inch.

Sound copper castings are difficult to produce due to the affinity to oxygen when melting. Melt on the oxidizing side with a charcoal cover and deoxidize with phos-copper 1½ to 2 ounces per 100 pounds. In the south we never melted and poured copper on a rainy day or one with high humidity. Large risers are required due to its high shrinkage. Dry sand molds, cement sand molds and Furan no-bake molds are your best bet for medium-to-heavy copper castings.

copper base alloy: Any alloy where the main constituent is copper. Copper is alloyed with lead, zinc, nickel, tin, silicon, etc.

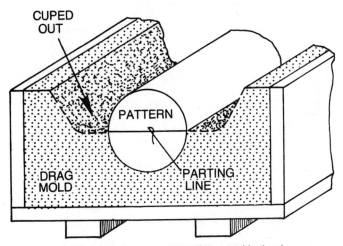

CUPED
OUT

PATTERN

DRAG
MOLD

PARTING
LINE

Fig. C-18. Coping out cuts away a portion of the sand in the drag.

85 percent copper, 5 percent tin, 5 percent lead and 5 percent zinc is the copper base alloy called *ounce metal* or *red brass*.

copper phosphorous: Also known as *phos copper*. It is a metal deoxidizer made of copper containing various percentages of phosphorous sold in shot and wafer form. The two most popular percentages are 10 percent and 15 percent. The phosphorous is combined with the copper to make it stable so it can be handled safely. The usual amount used in a red brass heat is 1 to 2 ounces. 15 percent Phos Cu per 100 pounds of metal (15 percent phos 85 percent Cu).

The copper melts, thereby releasing the phosphorous which combines with the copper oxides in the melt and releases the copper as oxide-free copper.

core: A pre-formed baked sand or green sand aggregate inserted in a mold to shape the interior part of a casting which cannot be shaped by the pattern (Fig. C-19).

When a pattern requires a core, a projection must be made on the pattern. This projection forms an impression in the sand of the mold in which to locate the core and hold it during the casting. These projections are called *core prints* and are part of the pattern.

Sometimes it is possible to make a pattern in such a way that a core will remain in the sand when the pattern is removed. The pattern for a simple shoring washer is made in this way.

core assembly: When cores are too complicated to produce with

227

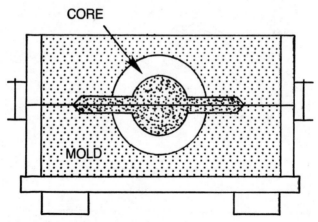

Fig. C-19. A core is often inserted in the mold.

ease as a single unit (core), a core assembly is made. It is made up of a series of cores (parts) and assembled by pasting, leading, bolting, etc. An example of this is the coring for a water-cooled engine block which is quite complicated but consists of an assembly of simpler cores.

core binders: Materials used alone or in combination to bind the sand grains together to form a core which are set by chemical action, heat, microwaves, etc. Linseed oil, fish oil, waterglass, urea, wheat flour, cement, dextrin, molasses, hydrol, sulfite, latex, furan, pitch, etc.

core box: Wood, metal or plastic, structure which has the shape of the desired core to be produced—the female shape of the desired core (male).

core box drawing machine: A machine which mechanically draws the rolled over core box from the core. There is a wide variety of machines built. Some roll the box plate and core over and draw the box upward, and some roll over and the core plate is lowered from the box. Some are completely automatic where they blow the core in the box and draw it. Others are semi-automatic, while some are operated completely manually.

core daubing: Also called *core mudding*. The joints of glued cores are daubed or mudded at these joints to make a smooth joint and to prevent metal from finning at the junction. Like paste, daubing can be purchased or homemade.

Red talc or graphite moistened with water to a soft mud works well, as does 3 percent bentonite, 3 percent dextrin or molasses and 94 percent silica flour mixed with water to a soft mud.

core driers: Metal plates that have the shape of the core in order to support it and keep it in shape during the baking cycle (Fig. C-20).

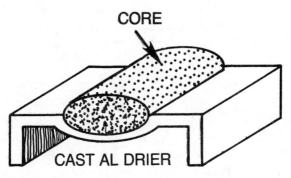

Fig. C-20. Core drier.

core, green sand: A core made of molding sand (not baked) with a core box formed by the pattern during molding and used in the green state (Fig. C-21).

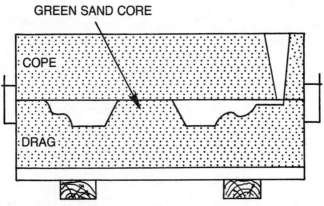

Fig. C-21. Green sand core.

core grinding: With close work such as engine cores, where precision is needed, the cores are made oversize at the joint and ground to the height desired (Fig. C-22). The core grinder consists of a table on which the core holding fixture is attached and an adjustable swing arm grinding wheel, which when adjusted to the desired height, is swung across the core in the jig grinding it to size and at the same time producing a true and level surface.

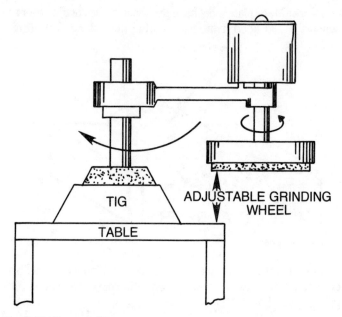

Fig. C-22. Core grinding.

core hardness: The measure of hardness with a core hardness scratch gauge (Fig. C-23). A spring loaded dial indicator device similar to a mold hardness gauge. It measures the depth of penetration of a steel knife edge as it is drawn across the core

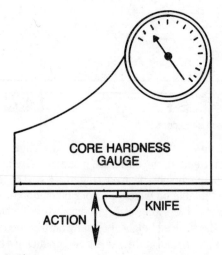

Fig. C-23. Core hardness.

with the instrument pressed against the core. The shoe rests on the core. Typical Readings:

- Dry Sand Molds:Soft—20
- Baked Cores:Soft—35
- Baked Cores:Hard—75
- Dry Sand Molds:Hard—40
- Baked Cores:Medium—50
- Baked Cores:Very Hard—90

core hooks: Hooks or pat eyes placed in a core used to lift or handle the core (Fig. C-24).

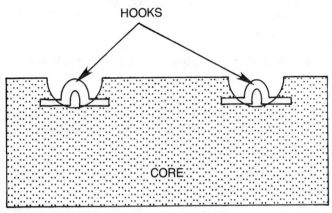

Fig. C-24. Core hooks.

core maker's trowel: The core maker's trowel is exactly like a finishing trowel with the exception that the blade is parallel its entire length and the nose is perfectly square (Fig. C-25). They come in widths of from 1 inch to 2 inches in ¼-inch steps and blade lengths of 4½ inches long to 7 inches long in ½-inch steps. This trowel is used to strike off core boxes and due to it being parallel and square ended, a core can be easily trimmed or repaired and sides can be squared up.

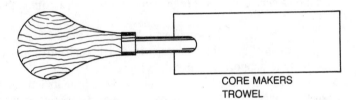

Fig. C-25. Core maker's trowel.

core marker: The print on a core which is so shaped that it can be set in the mold only one way matching a corresponding shaped print on the pattern (Fig. C-26). Also known as a *keeper* or *key*.

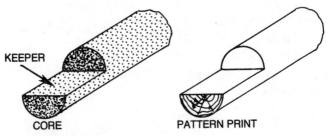

Fig. C-26. Core marker.

core molds: A mold made entirely of cores as the no-bake molding system is essentially core molds.

core paste: A prepared adhesive used to glue and join sections of cores together. You can make your own from waterglass and talc, dextrin, flour, graphite and molasses; however, it is more economical to purchase a prepared paste from a foundry supplier. There is a hot paste which is sold in stick form and applied with an electric paste gun which heats the paste and applies it with a trigger action.

core plug: This is a wood or metal form that is an exact reproduction of a desired core and its print (Fig. C-27). The purpose is to see exactly what the core will look like to determine the construction of the core box, boxes or driers.

In many cases the core plug is used to produce a plaster cast core box, then lagged to produce a pattern. This is a good quick method where only a few castings are required or a prototype piece.

core print: An extension on a pattern which makes a print in the mold cavity used to hold and locate the core in the proper position and alignment in the mold during casting (Fig. C-28). The core print on a core is that part of the core that rests in a print in the mold.

The so-called soft plugs or freeze plugs on the side of your auto engine are in reality where cores extended through into prints in the mold to support the core during pouring.

core rise: This defect is caused by a core rising from its intended position toward the cope surface. It causes a variation in wall thickness or if touching the cope, there is no metal at that point.

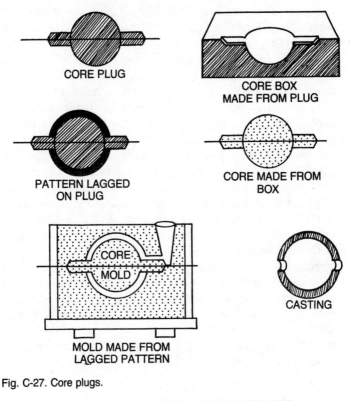

Fig. C-27. Core plugs.

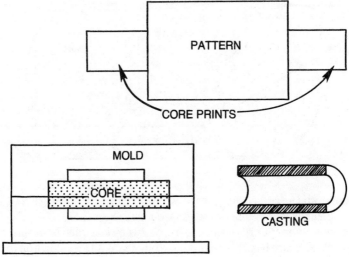

Fig. C-28. Core print.

The core has shifted from its position. A *green sand core rise* is when a green sand core in the drag is cracked at its base (caused when drawing the pattern) and floats toward the cope. Dry sand cores will float if the unsupported span of a thin insufficiently rodded core is too great. It will bend upward by the buoyancy of the metal. Insufficient core prints in number and design, insufficient chaplets, slipped chaplet, chaplets left out by molder or poor design of the core. This defect is easy to spot and remedy.

core rods: Rods and or wires incorporated in the core during the coremaking. They are used to give the necessary strength to the core to prevent it from breaking from the forces it is subjected to during the casting operation. It can be a small wire in some cases or large rods and wires in combination such as one would use to reinforce concrete in building a wall, etc. The size, amount and complexity are determined by the size of the casting and design of the cored cavity.

core sand: Sand free from clay, a nearly pure silica, any sharp sea sand used to make dry sand cores. Core sand is usually purchased as washed dried and graded. When core sand is bonded with clay or bentonite, tempered and used to mold with, it becomes molding sand.

core shift: Distorted or variation of a cored cavity in a casting due to the core moving (shifting) position, or the misalignment of cores in assembling (Fig. C-29).

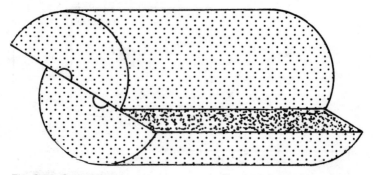

Fig. C-29. Core shift.

core sizing: Checking a core with a templet or other gauge to determine its go-no-go tolerance limits (Fig. C-30).

core sweeps: Cores produced from a sweep or sweep box in place of a full core box. Large cores are often swept to cut down on the pattern cost (core boxes) (Fig. C-31).

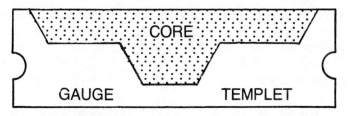

Fig. C-30. Core sizing.

core venting: Since more often than not the core is almost totally surrounded by metal, the natural venting properties of the core (sand) must be supplemented by creating additional passages leading from the interior of the core through the prints to the outside of the mold.

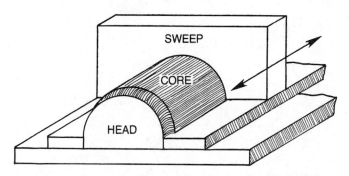

Fig. C-31. Core sweep.

core vent wax: A specially formulated wax used to vent dry sand cores (Fig. C-32). The wax is rammed in place during the coremaking and melts during the baking of the core, leaving a clean hole or vent in its place. It is sold in spools in sizes from 1/32 to ⅝ round solid, from ⅛ to ⅜ round hollow and 1/32 × ⅛ to ¼ × ¾ flat oval.

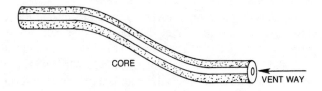

Fig. C-32. Core vent wax.

cores baking dielectric: Baking cores bonded with urea binders by passing them through dielectric oven (microwave oven). Typical core mix suitable for dielectric baking: 500 pounds sharp sand; 11 pounds urea formaldehyde base binder; 4½ pounds cereal; 1¼ pounds boric acid; 2½ quarts kerosene; and 8 quarts water.

corrosion: The oxidization of a material or matter such as rusting which is increased with the presence of heat and oxygen or materials which contain oxygen such as H^2SO_4 sulphuric acid.

cover core: A core that essentially covers the entire mold cavity with the exception of the gate entrance and is held down by the cope and in place by its print (Fig. C-33). In many cases the cope can be eliminated and a weight set on the cover core. A core sand pouring cup should be used to pour the casting—a great saving in labor.

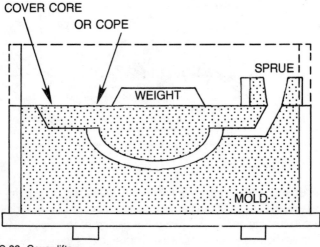

Fig. C-33. Cover lifts.

cover lifts: Tin forms used to cast the lifter bridge on stove covers (Fig. C-34). However, the clever engineer or inventor can find many other uses for them.

The tin is dropped into a socket in the pattern and the sand is sucked under the tin. Then the mold is finished. When the pattern is removed, the tin remains in the mold. The metal flows through the inside of the tin forming the crossbar or lifter bridge, performing a task which would be difficult to do with a core.

cracked molds: Molds that are cracked due to improper handling—flasks too weak, weak and rotten bottom boards,

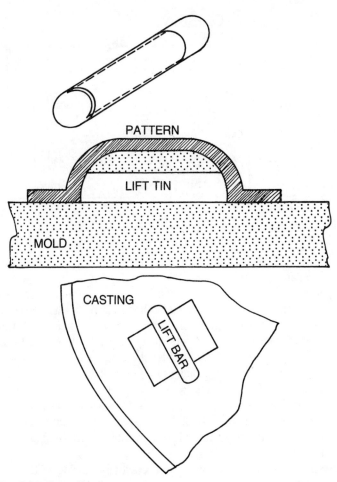

PATTERN

LIFT TIN

MOLD

CASTING

LIFT BAR

Fig. C-34. Cover lifts.

mold sitting on an uneven surface, loose flask bars. The defects caused by cracked molds are scabs, run outs and finning.

creep: The slow or gradual distortion of metal under a constant stress.

cribbing: Also called curbing (Fig. C-35). A large mold is surrounded with heavy planks or steel beams and the space between the cribbing is rammed with floor sand to back and strengthen the mold. In solid investment molds this practice is widespread due to the fragile characteristics of a calcined mold. Cribbing will also prevent a run out should the mold crack during pouring. Should the mold be lowered into a pit and then sup-

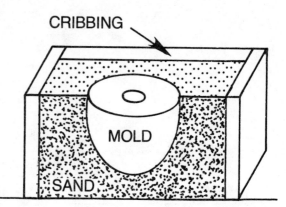

CRIBBING

MOLD

SAND

FLOOR

Fig. C-35. Cribbing.

ported by rammed sand, the term is *pitting*. The end results are the same.

crucible: A large family of ceramic pots or receptacles made of graphite and clay or clay and other materials such as silicon carbide, etc. Used primarily in the foundry to melt steel, iron, bronze, aluminum, etc. Pots made of iron, steel and wrought steel when used to melt-in are also referred to as crucibles. Crucibles cover a size range of a few ounces capacity to 2395 pounds (A #700) and up. Foundry crucibles are standardized by size and number. The rule of thumb is a crucible's capacity is one times its size number in aluminum and three times in red brass. A number 10 crucible will hold 10 pounds of molten aluminum or 30 pounds of molten brass.

crush: This defect is caused by the actual crushing of the mold causing indentations in the casting surface. It is caused by flask equipment, such as bottom boards or cores that are too tall or too large for the prints or jackets. Also caused by rough handling.

crush strip: A strip around a large pattern which will prevent the matching edges of a mold cavity from crushing when they meet (Fig. C-36). An intentional flash around the casting. Used on large castings where the cope due to its weight could twist or sag, preventing the cope from coming down on the drag evenly at all points.

The crush strip is removed from the casting by chipping or grinding. Any damage done by crushing will be confined away from the casting and not into the cavity.

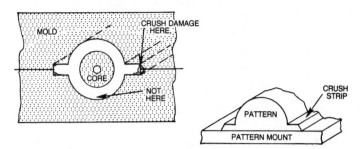

Fig. C-36. Crush strip.

crystobalite: A form of quartz with a melting point of 3140°F crushed and ground to flour. It is widely used in prepared investments.

cupola: A continuous melting device primarily for cast iron (but not limited to) which is in fact and operation a small type of blast furnace lined with refractories for melting the metal in direct contact with the fuel (coke or charcoal) by forcing blast air under pressure through openings (tuyeres) near its base. The cupola resembles a smoke stack and is often called a *stack*.

cupping: The tendency of tangential sawed pattern lumber to curl away from the heart of the tree (Fig. C-37). This is one reason why you should buy ¼ sawed pattern lumber.

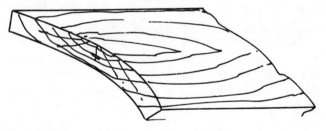

Fig. C-37. Cupping.

cut: Defects in a casting caused by the erosion of the sand by the molten metal flowing over the mold or cored surface.

cutting over: Working over floor or heap sand with a molders shovel to distribute the moisture, areate and otherwise obtain a uniform mixture.

cutting shellac: Diluting shellac with alcohol to the desired consistency. Dissolving flake shellac with alcohol. Always add a pinch of *oxalic acid* to cut shellac to prevent it from oxidizing and turning black.

dam gate: A gate constructed with a dam to effect a ready choke and skimmer (Fig. D-1). It is usually made with a triangular core which sets in a suitable print.

damper screw shells: Screw shells are drawn tin shells with a roller thread used for casting a threaded hole (Fig. D-2). They eliminate the cost of drilling and tapping. Since these threads are rolled, they are not satisfactory for a pressure-tight fitting.

The shell is dropped into a hole in the pattern before molding and the inside fills with green sand when the mold is rammed. The hole in the pattern must be large enough for the shell to draw freely. They can also be used on a core to produce a threaded hole.

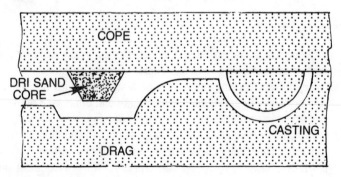

Fig. D-1. Dam gate.

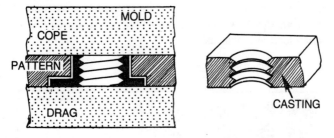

Fig. D-2. Damper screw shell.

daub: Refractory material used to build back and replace the burned back section of the cupola at the melting zone. It consists of fire clay and ganister or a commercial product such as plastic fire brick. After each run of the cupola, this area must be chipped back to a fresh slag-free lining, and the cupola daubed back to its original diameter at the melting zone. Also used to repair ladles, lips and furnace linings. The area which requires the repair must be thoroughly cleaned and wetted down and the daubing material rammed firmly in place. The daubing mix must be only wet enough to work. Excessive moisture will cause the daubing to steam and blow off when the bed is ignited or when the cupola is put under blast. Daubing repairs should be dried with a torch or in a core oven. If wet, they will fail or gas the metal in the ladle. It must be done correctly and carefully without rushing.

decarbonization: The loss of carbon from the surface of a ferrous alloy as the result of heating in a medium which reacts with the carbon.

decrepitation: The property of a material or substance to fly apart on being heated (explode).

deformation (sand): The change of the linear dimension of a sand mixture when put under stress.

degassing: The removal of absorbed gasses from a metal bath by chemical or mechanical action. Deoxidizing red brass with phosphorous copper would be a chemical action. Removing gas from an aluminum bath by bubbling dry nitrogen through it is a mechanical action as the nitrogen has no chemical action with the aluminum or dissolved gasses but creates a mechanical vehicle (bubbles) which the gasses attach themselves to and ride to the surface and out.

dentrite: The crystals which form during solidifcation (Fig. D-3). As a casting starts to freeze in the mold, the metal in immediate contact with the walls of the mold will freeze first, forming a thin layer of equi-axed crystals. Dentrite crystals then grow or shoot out from this face into the remaining liquid, forming crystals which look like pine trees.

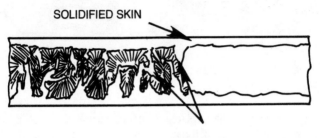

Fig. D-3. Dentrite.

deoxidize: The removal of the oxygen from metal oxides in a bath. When you rob them of their oxygen, they then revert back to metal. Has to be done by introducing a substance which has a greater affinity for oxygen than the substance which is oxidized (combined with oxygen) such as lithium or phosphorous for copper, aluminum for steel.

density: The mass of a unit volume of a material at different temperatures, measured as specific gravity.

desulphurizing compounds: Compounds used to remove or lower the sulphur in molten cast iron prior to pouring. Sulphur is primarily an undesirable element in iron.

The active desulphurizing agent in both caustic soda (76 percent) and soda ash (58 percent) is sodium oxide. Eighty to 85 percent or more of the sulphur present in cupola iron can be removed by a single ladle treatment.

Of the two reagents used to remove sulphur, sodium carbonate (soda ash) is the most used. A good commercially pure soda ash is best and is sold under various trade names.

desulphurizing ladle: A U-shaped ladle usually mounted in front of the cupola on trunnions with a tilting mechanism used to treat the iron with soda ash to remove the sulphur (Fig. D-4). The iron is tapped into the ladle, the stream of iron from the cupola spout going into the ladle through a slot shaped opening on one end of its insulated cover and then treated. Iron is poured into the receiving ladles from a tea pot spout on the opposite end by tilting the U ladle. The slot arrangement gives access to the iron

242

coming from the cupola regardless of the angle of tilt. Some loss of temperature results but is usually offset by the increase in fluidity of the desulphurized iron.

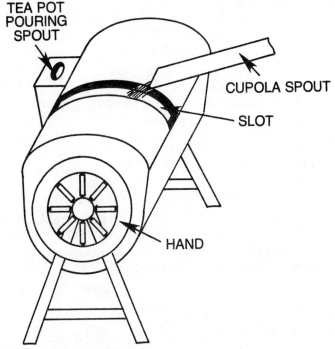

TEA POT POURING SPOUT

CUPOLA SPOUT

SLOT

HAND

Fig. D-4. Desulphurizing ladle.

dextrin: Also called *amylin*. It is a compound having strong adhesive properties, a white powder, odorless with a sweetish taste, water soluble. It is made by the action of dilute nitric and hydrochloric acids on starch, then heating to 100 to 125°C. It is used as a binder in skin dried, dry sand molds, core washes, etc.

dezincification: The loss of zinc from the surface of a zinc-bearing alloy such as brass or zinc corrosion. It is accompanied by the deposit of a residue of one or more active components—usually copper.

diffusion: The movement of atoms within a solution in a direction from a region of high concentration to one of low concentration in order to achieve homogenity of a solution which may be a liquid, solid or gas.

die casting: Casting into a permanent mold (metal or graphite) by forcing the metal into the mold cavity under high pressure. A rapid method of casting which is highly mechanized and only used where a large volume of castings are needed due to the high cost of dies and machines. It is limited by design and metal selection. The most used metal in this case is zinc and zinc alloys.

dirty castings: Dirty castings can come from many causes; however, most are caused by poor pouring practices, gating and sloppy molding. Additional cause include:

- Not skimming properly.
- Gates too large to choke properly.
- Not establishing and maintaining a choke.
- Mold cavity not blown out completely and properly.
- Excessive use of dry parting.
- Weak sand.
- Soft molding.
- Wild metal.
- Carbon too high for casting section (in iron).
- Gates nozzling (spraying) the metal into the cavity.

discs: Chaplet shims, plain or perforated tin plates used to provide a shim or to provide an additional bearing surface to rest the chaplet on in the case where the core is very heavy or subjected to excessive lift when the mold is poured (Fig. D-5).

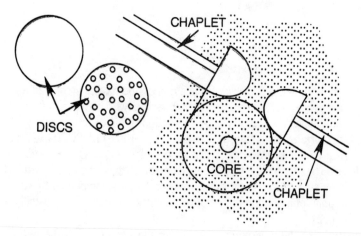

Fig. D-5. Discs.

On the cope side of the core the plates are simply layed in position prior to setting the cope. On the drag side they are usually pasted to the core with core paste. The paste is either oven or torch dried.

dispersion: Separation or scattering of fine particles in a liquid medium (deflocculation) used in connection with the fineness test of clay.

dissolved gasses: Gasses dissolved by a metal when heated and or melted. In most cases they come from the products of combustion and must be removed by the proper degassing treatment.

dolomite: A type of limestone used in making cement and lime, used as a flux in melting iron in the cupola.

double head chaplets: Mild steel double headed chaplets, tinned or copper plated, round or square heads.

dough roll: A roll made of flour and molasses water and rolled out on a flat board, like clay snakes or large macaroni (Fig. D-6). The dough roll is placed on the drag around the cavity prior to closing the mold. It acts as a gasket or seal to prevent a run out during pouring by taking up any unevenness between the cope and the drag mating surfaces.

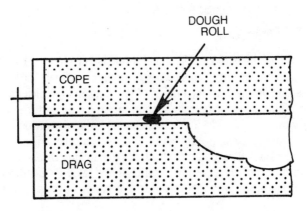

Fig. D-6. Dough roll.

dovetails: Tin forms which permit the accurate casting of dovetailed grooves either straight or tapered (Fig. D-7).

A brass lock strip is attached to the pattern which sticks up through a groove in the tin in order to locate it and hold it in position during molding.

dowel: A pin of various types used to join sections of parted patterns and core boxes to assure the correct registry.

draft: Pattern draft is defined as the taper on vertical elements in a pattern which allows easy withdrawal of the pattern from the mold (Fig. D-8). The amount of draft required will vary with the

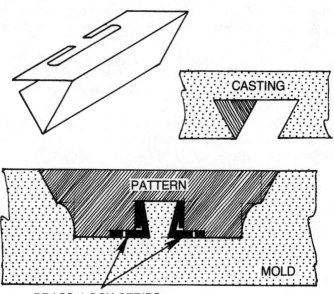

Fig. D-7. Dovetails.

depth of the pattern. The general rule is ⅛ inch taper to the foot which comes out to about 1 degree and on shallow patterns such as our disc a 1/16 inch taper of 0.5 degree is sufficient.

drag: The lower or bottom section of a mold or pattern.

draw back: A section of a mold that must be lifted away on an arbor or plate to facilitate the removal of the pattern (Fig. D-9).

draw bolt: A threaded eye which fits a matching thread in a plate (*draw plate*). It is let into a pattern and used to lift (draw) the pattern from the mold or otherwise handle it (Fig. D-10).

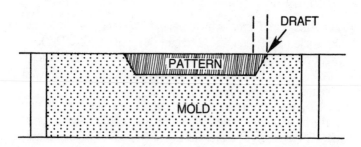

Fig. D-8. A draft.

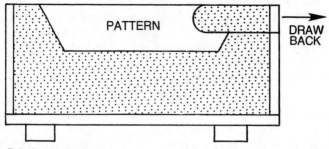

Fig. D-9. A draw back.

drawing: Removing the pattern from the mold. Removing a core box from core.

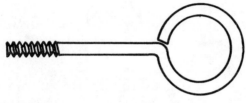

Fig. D-10. Draw bolt.

draw pins, screw and hooks: These items are used to remove or draw the pattern from the mold (Fig. D-11). The draw pin is driven into the wooden pattern and used to lift it out as a handle. On a short small pattern one pin in the center will do it. If the pattern is long, use one on each end for a two hand straight lift. The draw hook is used the same way. The draw screw is screwed into the pattern for a better purchase on heavier patterns and prevents the pattern from accidently coming loose

Fig. D-11. A. Draw screw; B. draw pin; and C. draw hook.

247

prematurely and falling, damaging the mold, pattern or both. In large patterns, plates are let into the pattern at its parting face which have a tapped hole into which a draw pin with a matching thread is screwed for lifting by hand or with a sling from a crane. Two or more are used.

Metal patterns, if loose, are drilled and tapped to receive a draw pin.

draw plate: A plate or device sometimes attached to the pattern to assist or facilitate the drawing of the pattern from the sand (Fig. D-12). A vibrating guide plate used to draw a core box from the core. In some cases it is a simple cutout in a plate through which the pattern is drawn.

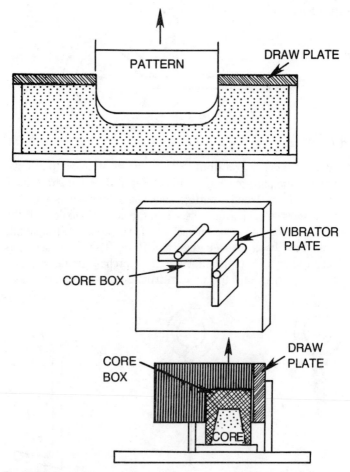

Fig. D-12. Draw plate.

draws: A shrinkage or draw at the point where the gate joins the casting (Fig. D-13). A common form of defect on chunky yellow brass castings. By adjusting the gating and filleting, it can usually be overcome.

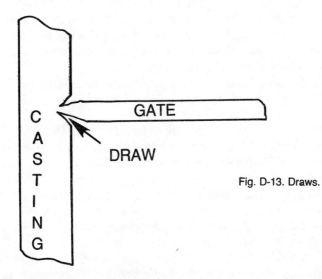

Fig. D-13. Draws.

draw stick: A pointed hardwood stick used to draw small light patterns from a mold.

driers (ladle): Heating devices used to dry ladle linings and pre-heat them (Fig. D-14). They can be electric or gas and oil fired. The lining must be dry and if the ladle is not heated high enough prior to tapping metal into it, a great loss of metal temperature will occur and the ladle will scull over or freeze.

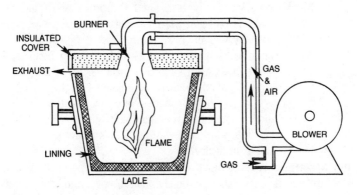

Fig. D-14. Driers dry ladle linings.

driers (sand): Any device used to dry wet sand fired with coal, gas, oil or electricity (Fig. D-15). Some are continuous driers. The sand tumbles through a heated and inclined revolving cylinder. Others have a simple coal or coke-fired pot belly stove arrangement with a drying jacked around its circumference.

drops: A portion of the cope sand drops into the mold cavity before or during pouring. The causes are bumping with weights, rough clamping, weak molding sand (low green strength) rough closing and jackets placed on roughly, etc.

dross: Metal oxides on a molten metal bath or as a defect found in and on the surface of a casting due to improper gating, melting practice or deoxidizing treatment.

dry binders: Binders used in core and mold making. They are dry. Includes a wide variety of prepared dry binders such as lion binder, truline binder, flour, rosin, wheat flour, corn flour, clay, pitch, goulac, etc.

dry sand molds: Any mold that has been dried (baked) prior to pouring. Core molds and Furan no-bake are also sometimes called dry sand molds. Although they are dry (no free moisture), they should be called by their respective names.

Dry sand molds are produced from green sand (system sand) or bonded crude silica sand or a combination of the two to which a baking type binder is added to provide a bond when the mold is dryed. This bond can be pitch, oil, cereal, etc., alone or in various combination.

The molds are placed (open) in a mold drying oven and dryed at 350 to 450°F overnight. They are allowed to cool, cored, closed up and poured.

Typical dry sand mixes:

- Heavy Brass—system sand tempered with glutrin water (1 pint glutrin to five gallons water), 5 percent pitch by weight, baked until dry.
- Medium to Heavy Grey Iron—60 gallons dry heap sand (floor or system sand); 30 gallons crude dry silica sand; 4 gallons bentonite; 3 gallons pitch.
- Heavy Phosphour Bronze—Naturally bonded molding sand, 25 gallons; river sand, 10 gallons; wheat flour, 1 gallon; blue clay water, 30 to 40 baume. Bake at 450°F until dry.

dry strength: A mold must not only hold it's shape in the green state, but it must also hold it's shape in the dry state. This is an important property and is measured as dry compression by

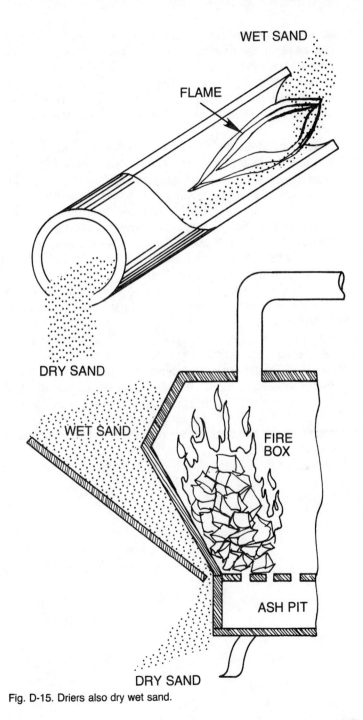

Fig. D-15. Driers also dry wet sand.

251

allowing the test specimen to dry out before testing which is then carried out in the same manner as for green strength. A good average figure is 30 pounds per square inch. Dry strength should be no higher than necessary. Excessive dry strength results in a critical sand. If the molding sand has a too high dry strength it will not give or break down as the casting shrinks during solidification.

ductile iron: Also called *nodular iron.* It is a cast iron where the graphite distributed through a ferritic matrix is spheroid or nodular in shape instead of flat flakes in a matrix of pearlite and ferrite. It is produced by adding (treating) the molten iron with an alloy of magnesium, copper-ferrosilicon or a nodulizing alloy marketed by various firms for the introduction of magnesium into iron. The treatment leaves up to .08 percent residual magnesium in the iron. The free graphite in ductile iron in spheroid (ball) shapes gives the iron its ductile characteristics.

ductility: The measure of the degree of permanent deformation (elongation) a metal or material will take before fracturing or breaking.

dune sand: Wind blown deposits of sand—90 percent silica, usually located near large bodies of water, but not necessarily.

durability: The measure of the sands ability to withstand repeated usage without losing its properties and to recover it's bond strength after repeated usage. The sands fineness and the type and amount of clay bond determines the sands durability. The ability of the bonding clay to retain it's moisture is also an important factor.

dust bag: Pourous cotton cloth bags used to apply dry parting and facing materials to a sand mold.

E

electric riddle: An electrically-powered riddle. See also gyratory riddle.

electrodes: Welding rods used for electric welding and carbon rods used in the arc furnace for melting.

elongation: The amount of extension in the vicinity of the fracture in the tensile test. It is expressed in percents of the gauge length 2 inches (Fig. E-1). The coupon sample used to test the tensile strength is first marked off with two center punch marks 2 inches apart. The coupon is pulled (stretched) until it breaks on the tensile strength machine, giving the tensile strength. The two ends of the sample are then fitted together at the point of fracture and the distance between the punch marks measured to determine the percent of elongation.

ethyl silicate: A silicic acid used as the principal binder for ceramic investment molds. The normal content of ethyl silicate is 25 percent available silica. Water hydrolyzes ethyl silicate to alcohol and silicic acid which dehydrates to an adhesive amorphous silica.

etching: Attacking a highly-polished metal sample surface with a weak acid or other corrosive agent (caustic soda for aluminum). The etching agent attacks the various constituents of the metal unequally so that when the etchant is removed by washing and drying, it can be examined under a microscope or in some cases by eye.

eutectic: The alloy structure of two or more solid phases formed from a liquid eutectically.

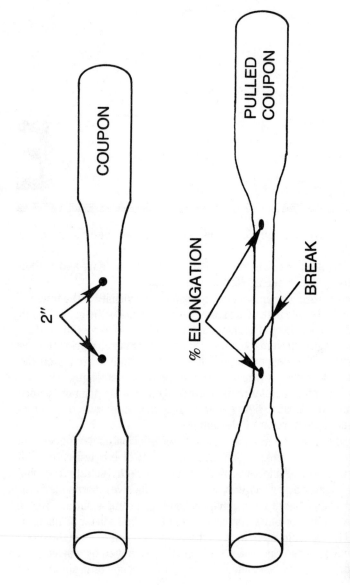

COUPON

2"

% ELONGATION

PULLED
COUPON

BREAK

Fig. E-1. Elongation is the amount of extension in the vicinity of the fracture.

A pure metal solidification temperature may be lowered by adding another metal. The reduction of the solidification temperature is proportional to the amount added up to a point. The solidification range of a high copper alloy such as 95 percent copper would run somewhere in the neighborhood of 1850°F to 1700°F against 1600 to 1580°F for a similar alloy with only 58 percent copper.

If we add sodium chloride (salt) to water we lower the freezing point of the water. Then as the freezing point is reached the ice crystals (salt free) forming make the remaining unfrozen portion higher in salt content, reducing the freezing point still further down the scale. When we reach this point by lowering the temperature when ice crystals start to form again, the same thing happens until we finally reach the point where the water and salt freezes. This final frozen mass of frozen ice and salt crystals is the eutectic. The *eutectic temperature* is the point where this freezing happens.

It is the basis of equipment used today to determine the carbon, silicon and total carbon. You pour a sample of molten iron in a special cup containing a thermocouple and plot the solidification curve. The curve is displayed on a digital LED read out in percents of carbon, silicon and total carbon of cast iron. This method is called *thermal analysis*. It is the measurement of the cooling curve of a solidifying iron sample and the electronic interpretation of the curve in terms of carbon, silicon and total carbon.

exoband: A sodium silicate (waterglass) binder used in the carbon dioxide mold and core bonding system.

expansion: The dilation or increase of size that sand or matter undergoes when heated.

expendable pattern: As in lost wax casting the pattern is lost. Expendable patterns for sand casting are styrofoam and shaped to the desired form with attached styrofoam gates, runners and risers. The styrofoam pattern is molded with dry, clay-free sharp silica sand in a box or steel frame. The pattern is vaporized by the metal poured into the mold, leaving the casting.

F

facing: The layer of sand surrounding the mold cavity. When a mold is made whereby using a specially prepared molding sand in the mold adjacent to the pattern to produce a smoother casting surface or a more refractory surface for the metal to lie against. It is then backed with system or floor sand. It can be simply new or finer sand or sand containing a bond such as flour or pitch.

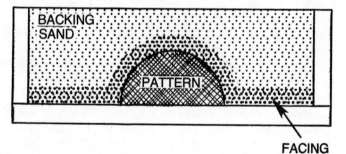

Fig. F-1. Facing sand.

false cheek: Making a mold from a pattern which would require a cheek only doing it in a two part flask by double rolling the mold (Fig. F-2). The cope half of the mold is made first, rolled over and the portion of sand which would normally be in the cheek is formed. The drag is rammed against the cope and false cheek, removed and the drag half of the pattern is then drawn. The drag replaces the false cheek. The mold is then rolled over and the cope removed. The cope section of the pattern is removed from the false cheek and the mold is closed.

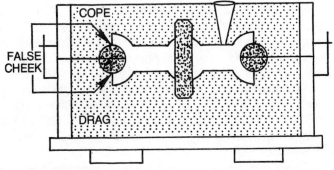

Fig. F-2. False cheek.

false core: A core made of green or dry sand to take care of an undercut on a pattern and removed and glued or wired in position in the mold proper.

The core is made by tucking the sand into the under cut (back draft) and forming a print on the core. When the mold is rammed the false core will leave a corresponding female print into which it is secured.

fasteners, corrugated: Corrugated (nails) fasteners used in pattern construction to secure stock together when turning a split pattern (Fig. F-3).

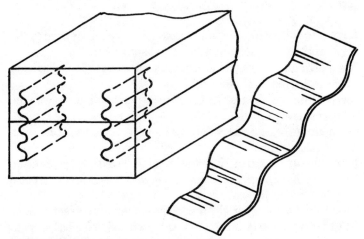

Fig. F-3. Corrugated fasteners.

257

fatigue: The tendency for a metal to crack or break under conditions of repeated cyclic stressing. Bending back and forth, vibrating, continuously working.

feeder: A reservoir of molten metal to make up for the contraction of metal as it solidifies. To prevent shrinkage cavities in the casting (voids lack of feed metal) as a riser or head.

feeding: Pouring additional molten metal into a freshly poured mold to compensate for volume shrinkage while the casting is setting. Also referred to as *topping off* or *hot topping*.

feel of molding sand: Tempering a molding sand to the point that it feels correct. Regardless of the amount of instrument checks a good feeling sand usually rams easier and better. This skill of telling when a molding sand is right comes only with experience and practice.

ferrite: Pure iron consists of crystals known as ferrite which are only slightly attached by the usual etching agents except at the grain boundaries.

ferro alloys: Alloys of iron and another element or elements (other than carbon) when used as a raw material in the manufacture of ferrous metals.

ferro boron: An alloy that is 10 percent boron and 90 percent iron.

ferro chromium: An alloy of iron and chromium containing from 66 to 72 percent chromium and .06 to 7 percent carbon. Used to add chromium to a metal bath.

ferro manganese: Iron manganese alloy containing over 30 percent manganese used to introduce manganese in a metal bath.

ferro molybdenum: An alloy that is 58 to 64 percent molybdenum and the balance iron.

ferro phosphorous: An alloy of iron and phosphorous used to add phosphorous to steel to make phosphorous steel.

ferro silicon: An alloy of iron and silicon used for the addition of silicon in iron and steel baths.

ferro titanium: An alloy that is 17 to 38 percent titanium with the balance iron.

ferro vanadium: An alloy that is 35 to 40 percent vanadium, with the balance iron.

ferrostatic pressure: The pressure of the liquid steel poured into a mold. As long as the metal remains liquid this pressure is equal in all directions.

ferrous: Alloys in which the predominant metal or solvent is iron.

fillet irons: Steel or brass tools with each end being ball shaped. The ball is used to apply wax fillets to patterns (Fig. F-4). They are sold singularly or in sets from 1/16 inch ball to ⅝ inch ball. Each tool consists of two different size balls. A number 2 tool would have a ¼-inch ball on one end and a 3/16-inch ball on the other end. The tool is warmed over an alcohol lamp and the fillet is rubbed into place, using the tool with the radius corresponding to the selected fillet.

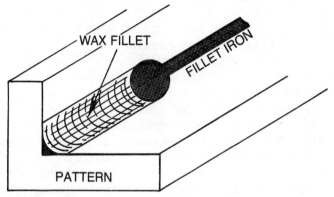

Fig. F-4. Fillet iron.

fillets: Patternmakers fillets are made of wax, wood, paper or leather and are used to form a radius on a pattern or core box (Fig. F-5). Wax fillets are applied with a fillet iron. Paper, wood and leather fillets are glued in place. The choice of which type of fillets to use depends upon the end use of the pattern. A pattern to be used only for a limited run now and again, wax is fine, or if it is a master from which metal production patterns are to be made, where if the pattern is large or to be used quite often, wood or

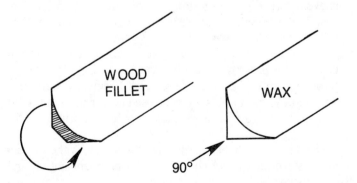

Fig. F-5. Fillets can be made of wax, wood, paper or leather.

leather might be the better choice. For small to medium patterns of jobbing work, 90 percent are filleted with wax. Paper fillets went out of use some time back but are available. Wood fillets can be used for straight runs only. They are cut slightly over 90° with the back relieved so that they will form a good tight fit when installed.

One-half to 1 inch leather fillets are shaped like wood fillets with a relief at the 90° junction. Wood fillets are, in some cases, glued and bradded into place. Never use brads on leather fillets as they draw the leather making an uneven surface.

When applying fillets to a pattern or core box there is a tendency to rush it—*don't*. You should apply the fillets with the same care as in building the pattern.

film on metal: When metal falling from a single gate has to climb a hill to fill the other side of the mold, often times a film is formed when the metal in the first valley stops momentarily then spills over into the second valley (Fig. F-6). This temporary halt will cause a reflected cold-shut and in some instances a sufficient film will form on the halted metal to separate part of the casting into layers. Two or more gates should be provided so that none of the metal has to rise over a hill and then run down into the valley on the other side. A gate leading into each valley will allow the mold to fill uniformly.

filters: A pad of coarse steel wool placed at the bottom of the sprue when casting magnesium to eliminate vortex action of the metal stream and assist in filtering out oxide skins (Fig. F-7). Slot sprues are used in casting magnesium to prevent vortex.

fin: This defect is a fin of metal on the casting caused by a crack in the cope or drag. It is caused by wracked flasks, bad jackets, uneven warped bottom boards, uneven strike off, insufficient cope or drag depth, bottom board not properly rubbed on drag mold sitting uneven (rocking).

fineness: A measure of the actual grain sizes of a sand mixture. It is made by passing a standard sample, usually 100 grams, through a series of graded sieves. About 10 different sieve sizes are used. As most sands are composed of a mixture of various size grains there is a distribution of sands remaining on the measuring sieves.

The fineness number assigned to the sample is the number of the sieve through which passed the largest amount of sand.

finger gates: Multiple gates into a casting from a single runner (Fig. F-8).

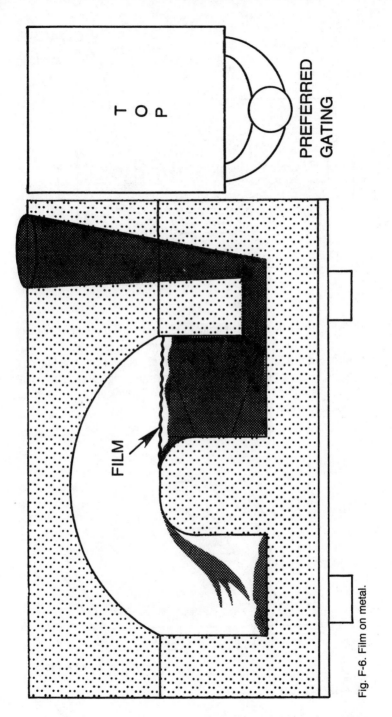

TOP

PREFERRED GATING

FILM

Fig. F-6. Film on metal.

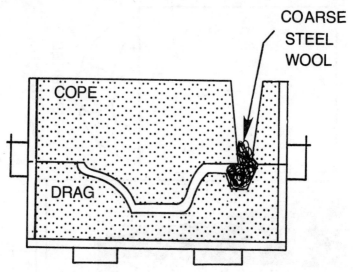

Fig. F-7. A filter.

finish allowance: The amount of extra stock left on the surface of a casting for machining plus shrinkage compensation.

finish mark: The symbol in the form of *f* appearing on a line of a print or drawing indicating that the marked edge of the surface of the casting is to be machined finished. It indicates the need for additional stock at the point or surface indicated.

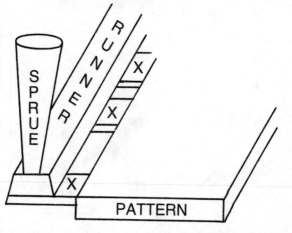

Fig. F-8. Finger gates.

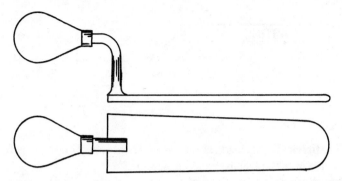

Fig. F-9. A finishing trowel.

finishing trowel: The finishing trowel is used for general trowel of the molding sand to sleek down a surface, to repair a surface or cut away the sand around the cope of a snap flask (Fig. F-9).

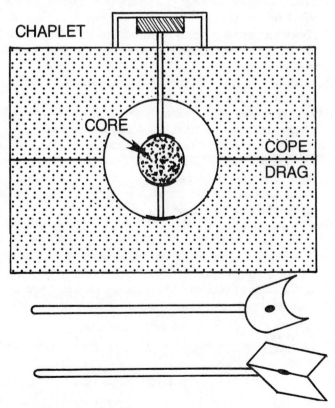

Fig. F-10. Fitted head chaplet.

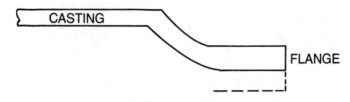

Fig. F-11. Flange.

fire brick: Bricks made of clay which fire at cone 27 or higher. For use in furnace linings and kilns.

fire clay: Any clay with a fusion point not less than P.C.E. cone 19, 2768°F up to cone 26 clays above P.C.E. cone 27 or higher are designated as refractories.

fitted head chaplets: A stem chaplet in which the head is fitted or shaped to fit a particular shape (on the core) (Fig. F-10).

flange: A stiffening member or a means of attachment to another object (Fig. F-11).

flat back patterns: Patterns that are flat on the side presented to the cope and do not extend into the cope of the mold. The entire mold cavity is in the drag (Fig. F-12).

flame annealing: The process of softening a metal by heat from a high temperature flame.

flame hardening: The application of heat to a metal from a high temperature flame to quenching (cooling as required).

flare point: The temperature of a metal such as yellow brass and high strength yellow brass (manganese bronze) where the zinc flares send little pips of zinc oxide from the metal surface. Low tensile strength manganese will flare at about 1850°F and high tensile Mg. Br. at or close to 1950°F.

flax swab: Also called a *horse tail,* it is used to swab the sand around the pattern at the junction of pattern and sand (Fig. F-13). To dampen this sand to prevent it from breaking away when the pattern is rapped and lifted from the sand, it is dampened by dipping it into a pail of water and shaking it out well. Used on floor work where the pattern presents a fair-sized perimeter.

Fig. F-12. Flat back patterns.

The swab is also used by some molders to apply wet mold wash or blacking to a mold surface. This requires great dexterity.

Fig. F-13. A flax swab.

flint shot: Flint rock ground, crushed and graded for use in place of sand for sand blasting. Flint nodules look like pearls and are free from cleavage lines. It is hard and clean and outlasts sand 3 to 1.

floor flasks: Flasks that are too large to be handled on the bench (Fig. F-14). These can be of wood or metal. The wood flasks are constructed in such a manner that the long sides provide the lifting handles for two man lifting and handling.

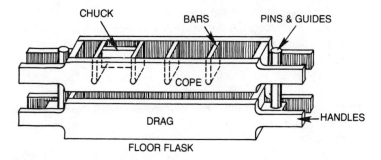

Fig. F-14. Floor flask.

In floor flasks, unlike bench flasks, when you reach a size of 18 inches × 18 inches to 30 inches × 30 inches, members are put in the cope which are called *bars*. These bars help support the weight of the sand in the cope to prevent it from dropping out. Bars are required in the drag half of the flask only when it is necessary to roll the job over and lift the drag instead of the cope in which case they are called *grids*.

The bars do not come all the way to the parting but clear the parting and portion of the pattern that is in the cope by a minimum of ½ inch. These bars in many cases have to be contoured to conform with the portion of pattern that is in the cope.

In all cases the bars are brought to a dull point along their lower edge to make it possible to tuck and ram the sand firmly under the bars.

The inside surfaces of both the flask and the bars in the cope section are often covered with large headed roofing nails with the head projecting ⅛ to ¼ of an inch. This gives the entire inner surface an excellent tooth and a good purchase on the sand.

Of course the floor flask like all others must be provided with suitable points and guides on both ends.

floor rammer: Same purpose and design as a bench rammer only the butt and peen ends are made of cast iron and attached to each end of a piece of pipe or a hickory handle. The average length is 42 inches.

flowability: The ability of a molding sand to flow and be easily rammed around a pattern. The ability of a sand to pack in and around a pattern with a uniform density.

flow rate: The relationship of pouring rate, height of sprue, area and diameter of the sprue. For example, a round sprue ¾ inch in diameter 5 inch tall will deliver 1 pound of molten aluminum per second. A sprue the same size, only 15 inches tall will deliver 1¾ pounds per second. A 2 inch sprue 5 inches tall will deliver 5¾ inches per second, and at 15 inches tall, 2 inches in diameter it will deliver 10 pounds per second.

fluidity: The ability of molten metal to flow readily as measured by the length of a standard spiral casting.

flushing of metal: Bubbling an inert gas through a metal bath with a metal or carbon lance to remove hydrogen, oxygen, etc., from the molten metal.

flux: Applying a solid or gaseous material to molten metal in order to remove oxide dross and other materials. Sometimes referred to as refining.

flux inclusions: Flux and or slag poured down the sprue and winding up on or in the casting. Caused by improper ladle or crucible skimming, not choking, improper gates or completely emptying a ladle or crucible during pouring.

fluorspar: A commercial grade of calcium fluoride CaF_2 used as a flux in ferrous melting.

fly ash: A finely divided product of combustion, sometimes found as inclusions in metal melted in an open hearth type of furnace.

foaming metal: Manganese bronze and aluminum bronze have a tendency to form like beer. This foam in turn solidifies and appears as dross in or under the surface of the casting.

The least agitation in handling causes foaming. The metal must enter the mold quietly at the bottom and fill the mold by simple displacement.

follow board: A board with a cavity or socket in it which conforms to the form of the pattern and defines the parting surface of the drag (Fig. F-15). It can be made of wood, plaster or metal. When made of sand it is called a *dry sand match.*

The pattern rests in the follow board while making up the drag half of the mold and in doing so establishes the correct sand parting. The follow board is removed leaving the pattern rammed in the drag up to the parting. The cope then takes the place of the follow board and is rammed in the usual manner. A simple follow board might consist of a molding board with a hole in it to allow the pattern to rest firmly on the board while the drag is rammed.

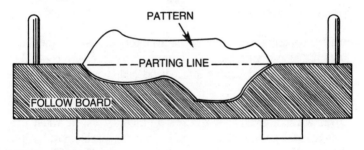

Fig. F-15. Follow board.

forehearth: A refractory lined reservoir in front of a cupola connected to the cupola by a trough. It is used to collect the metal from the cupola (Fig. F-16). The cupola runs continuously and the metal is tapped from the forehearth. It may be treated in the forehearth. A large quantity of metal can be accumulated in the forehearth for a large casting.

formaldehyde: Also called *methylene oxide.* It is produced by the oxidation of methyl alcohol. Its composition is HCHO.

By condensing *urea* with formaldehyde you produce *urea-formaldehyde* resins which are widely used in the foundry industry as binders for cores and shell molding (sand shell, not ceramic shell.)

foundry: An establishment or works where the production of castings is the principal endeavor.

foundry sprinklers: Sprinkler cans used by molders to add moisture to molding sand to keep it in condition.

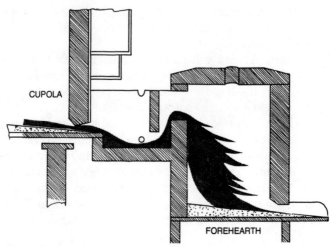

Fig. F-16. Forehearth.

foundry torches: Kerosene or oil-fired torches used in the foundry for skin drying molds. They produce an easily controlled flame. Also used to dry small hand ladles which have been daubed.

freezing range of alloys: This refers to the temperature range over which a given alloy freezes. In most cases the wider this range is the greater chance of segregation, a problem with some alloys. Silicon bronze solidifies over a range of 160°F, where copper 90 percent and tin 10 percent solidifies over a 300°F range. With copper 90 percent and zinc 10 percent, the range of solidification is negligible.

freezing ratio:

$$\frac{\dfrac{\text{Surface Area of the Casting}}{\text{Volume of the Casting}}}{\dfrac{\text{Surface Area of Riser}}{\text{Volume of the Riser}}}$$

If the two are equal, the freezing ratio is unity, and the riser and the casting will solidify at the same rate.

french gates: The system when pouring flat plates on end rather than horizontally. The french gate is attached to the casting its entire length (Fig. F-17).

french rammer: The french rammer is a piece of 1 inch diameter cold roll steel turned to an eye on one end and slightly tapered on the other (Fig. F-18). The bulk of the ramming is done with the eye.

268

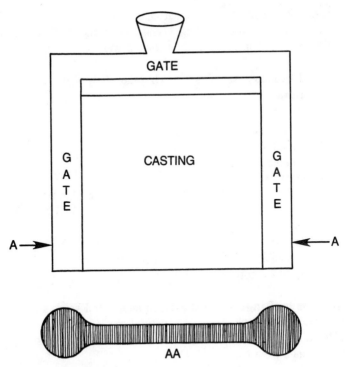

GATE

G
A
T
E

CASTING

G
A
T
E

A →   ← A

AA

Fig. F-17. French gates.

french sand: For many years French sand was considered the finest naturally bonded molding sand for fine art work, plaques, statuary, etc. Many of Rodan's pieces were cast in French sand. This sand comes from Fontenay des Roses near Paris. I have been told that they have been digging sand from this pit for some 400 plus years. It was common practice for many foundries to

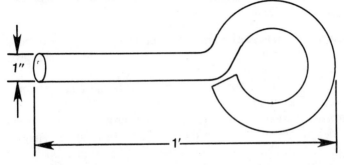

1″

1′

Fig. F-18. French rammer.

keep a ½ ton or more of French sand in the shop for that special job, or a gravemarker casting. Using it as a facing, riddled against the pattern ¼-inch thick, then backed with regular or system sand.

Of 10 random samples French sand showed the following test average:

- Clay, 17.2 percent
- Permeability, 18 at 7.2 percent moisture
- Grain Fineness, 176
- Green Strength, 10.5
- Shear, 3.1

Yankee sand which replaced French sand during World War 1 comes from around Albany, N.Y. An average of 10 samples showed:

- Clay, 19.8
- Permeability, 18.4 at 6.5 percent moisture
- Green Compression, 16.7
- Shear, 3.8

fuels for melting: Any fuel used to melt metal solid, gaseous, liquid or electrical, coal, coke, natural gas, oil, etc.

fuller's earth: A clay closely related to bentonite. Has also been used as a sand binder.

full pattern: A pattern that is constructed in one piece and not requiring that it be split in order to mold the job (Fig. F-19).

Fig. F-19. Full pattern.

fuming bronze: Flaring manganese bronze. It is believed by some that a few minutes of fuming, then replacing the lost zinc gives you higher tensile strength in the casting.

This school is very much divided. Some say under no conditions flare the heat.

furans: Various derivatives of furfural produced from formaldehyde and acetylene are known collectively as furans. They are used to produce resins and plastics. The furan most widely used in the foundry is furfural alcohol resins. These resins are the basic of the *Furan no-bake process* of molding and core making and has grown in leaps and bounds in the last several years because of the following advantages:

- Simple sand mix—resin, sand and activator.
- Excellent flowability.
- Reduced rodding.
- Reduction of skilled personnel.
- Less core and mold finishing.
- Uniform hardness.
- Closer tolerance and smoother casting.
- No ovens required, self curing.
- Easy shake out.
- Smaller molds due to its strength.
- Cores and molds do not develop thermal cracks when poured.
- Molds can be used for ferrous or nonferrous molding.
- No flasks required.
- Less weighting.

The normal formula for Furan no-bake system is 2 percent binder against the weight of the sand—100 pounds sand, 2 pounds furan binder. The activator or catalyst to set the binder is phosphoric acid (85 percent). The amount of binder required is 25 to 30 percent of the weight of the binder. The average formula is 100 pounds sharp sand, 2 pounds binder and ½ pound phosphoric acid (85 percent).

The activator is mixed with the dry sand, then the binder. The time the sand takes to set up is varied with the amount (percent) of activator used. Most pattern coatings are soluble in furfural alcohol. Coupled with the fact that heat is generated by the polymerization of the resin. You must use uncoated patterns, dry parting or wipe the pattern or core box with kerosene and rub the surface with graphite or mica. (The kerosene parting system works best). There is some commercial pattern coatings available for the no-bake system.

furnace (induction): The induction furnace operates by the process of induction (Fig. F-20). The metal charge is made the secondary of a high frequency transformer 1000 to 3000 cycles. Eddy currents are induced in the metal charge in much the same way a transformer introduces them in the secondary winding. The winding in this case (secondary) is the metal charge which acts like a short circuited secondary.

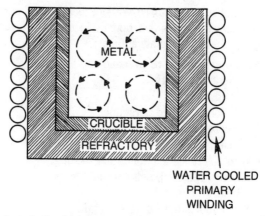

Fig. F-20. An induction furnace.

furnace (passive): A cope charcoal or coal-fired nonferrous furnace which depends upon its blast air for combustion by the draft created by a chimney or stack (Fig. F-21). The fuel and crucible resting on a grate.

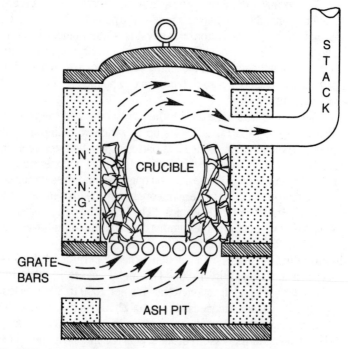

Fig. F-21. A passive furnace.

furnace, pit: A furnace usually placed in a pit with the cover at floor level.

furnace, tilting: A furnace which can be tipped in order to pour the molten metal bath into a receiving or pouring ladle by hand wheel, lever, hydrolic or air cylinders.

fused silica: Silica that has been fused at a high temperature and ground into various meshes (particle sizes). Used as a refractory and as a stucco for ceramic shell molds.

fusion: This defect is a rough glassy surface of fused melted sand on the casting surface either on the outside or on a cored surface. The cause is a too low sintering or melting point of the sand or core. It is quite common when a small diameter core runs through an exceptionally heavy section of great heat which actually melts the core. This can also be caused by pouring hotter than necessary for the sand or cores. A mold or core wash can prevent this in some cases but if the sintering point of your sand is too low for the class of work, you need a more refractory sand, zircon, etc.

gaggers: L-shaped metal rods square or twisted used to assist in the supporting of the sand in the cope of a floor or larger mold (Fig. G-1).

gainster: A highly siliceous sand stone used as a refractory, particularly for furnace linings, crushed and sized used as a grog in refractory clays.

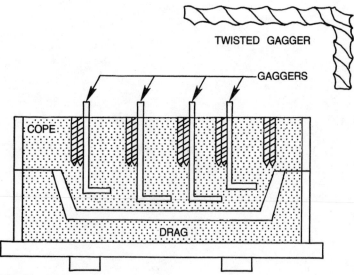

Fig. G-1. Gaggers.

garnet paper: Sand paper covered with crushed garnets in place of sand. A high grade sand paper preferred by patternmakers because of its long life and superior sanding ability.

gas generated: The gas generated by a core or mold when heated by the molten metal.

gas porosity: This defect is widely dispersed bright round holes which appear on fractured and machined surfaces. It is caused by gasses being absorbed in the metal during melting. The gas is released during solidification of the casting. The cause is poor melting practices (oxidizing conditions) and poor deoxidizing practices.

gate cutter: The best gate cutter is made from a section of a Prince Albert Tobacco Can (Fig. G-2).

To cut a gate or runner in green sand the tab is held between the thumb and first finger and the gate or runner is cut just as you would cut a groove or channel in wood with a gouge. The width is controlled by bending the cutters' sides in and out. The depth is controlled by the operator.

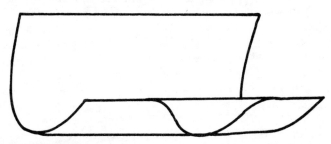

Fig. G-2. Gate cutter.

gated matchplate: A matchplate where the gates and runners are attached permanently to the plate and patterns (Fig. G-3). In a pressure cast matchplate, the gates, patterns and runners are all one casting (unit).

gated pattern: The gate is the channel or channels in a sand mold through which the molten metal enters the cavity left by the pattern (Fig. G-4). This channel can be made in two ways. One way is by cutting the channel or channels with a gate cutter. The other is by having the pattern have a projection attached to it which will form this gate or gates during the process of ramming up the mold.

gates and risers: The plumbing system used to fill a mold cavity (Fig. G-5).

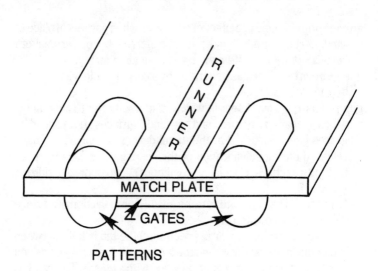

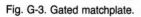

Fig. G-3. Gated matchplate.

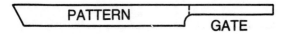

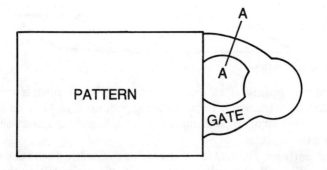

Fig. G-4. Gated patterns.

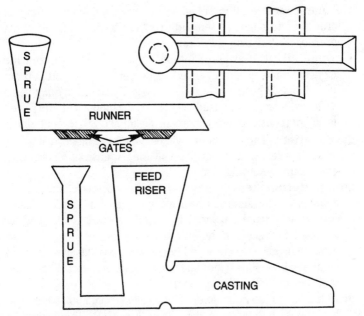

Fig. G-5. Gates and riser.

geared ladles: Large or small ladles which are equipped with a hand wheel and gearing mechanism used to tip the ladle.

geneva chaplets: A tin-plated sheet steel strong chaplet with a good bearing surface and unrestricted metal flow (Fig. G-6).

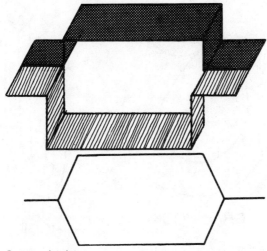

Fig. G-6. Geneva chaplets.

German crucible clay: A high grade clay used in the making of crucibles. It is sold bolted and used as a shake on facing and a print back. See also *bolted cement*.

German silver: A white alloy of copper, nickel, zinc and the highest grade of extra white: 50 percent copper; 30 percent nickel and 20 percent zinc. The cheap grade called *fifths* has a yellowish color and is used where the finished part to be plated is: 57 percent copper; 7 percent nickel and 36 percent zinc.

germination: The condition where you have abnormal grain growth. A few grains may grow at the expense of neighboring grains during solidification.

glue patternmakers: Glue used to glue up stock and pattern members. The standard was for years animal hide hot glue, cooked in a double boiler. However, it has been replaced by various liquid glues in the past several years.

glutrin: A lignin sulphite liquid core binder also used as a mold spray and in washes, glues and muds as binders, etc. When dried to a powder it is called *goulac*.

grab hook: A grab hook differs from a sling chain hook where the hook hooks between the chain links (Fig. G-7). The opening is much wider so that it can be hooked into the rollover pad eyes on large flasks. Also used to lift large copes.

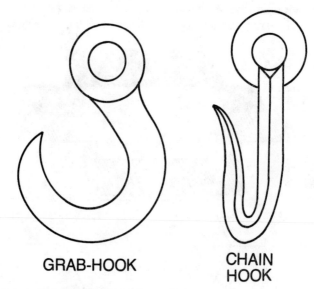

GRAB-HOOK   CHAIN HOOK

Fig. G-7. A grab hook and chain hook.

goulac: Goulac is a dry water soluble powder made from lignin sulphite liquor. It is a by-product of the paper pulp industry and used as a dry core binder in facing sand mixes. When mixed with water, it is used as a mold, core spray and wash binder, and in core pastes as dough rolls. It is a replacement for molasses in the foundry.

graphite: A crystalline or amorphous carbon used in the foundry as a mold coating and core wash. It is an additive to some molding sands and crucibles. It goes by many names including *blacking, plumbago, silver lead* and *ceylon lead*. By whatever name used, it is a form of carbon which can differ greatly in its physical properties. The finest grain is a crystalline graphite from ceylon. The chief action, when used as a facing for a sand mold, is that it fills in the spaces between the sand grains providing a smooth highly refractory surface upon which the metal can lie. This coating allows the sand to peel cleanly from the casting and prevents burning in—sand melting on to the casting surface. It is amorphous graphite when used for cast iron. A carbide iron surface is formed on the casting due to it being dissolved by the molten iron which produces a dirty hard surface. When crystalline graphite is used it is not dissolved and imparts a smooth easy-to-machine surface that is bluish in color.

Our first encounter with graphite is a lead pencil which is graphite (not lead). However lead will make a mark very much like graphite when rubbed on paper. This is probably how graphite got one of its many names including *black lead*.

As a wet mold and core wash, it is combined with a suitable binder and sold under many trade names. Some companies put out graphite washes which contain graphite with a binder which you mix with a volatile liquid as a vehicle. Alcohol and naptha are the most common. These washes are sprayed or brushed on to a mold or core surface and then ignited. The vehicle burns and the resulting heat sets the binder forming a firm baked on coating. When the vehicle is water with a water soluble binder such as linglin sulphite, the coated mold or core must be dried with a torch or in an oven to remove the water and set the wash. As a dry mold facing it can be used as is or with a small amount of wheat flour as a binder.

It is applied to the mold by shaking it on the mold cavity through a porous cloth bag *(blacking bag)* or brushed on with a fine camel hairbrush. The excess is blown out with the bellows.

It is also sold in graded flat plates or crystals used to form a cover on melting bronze to prevent oxidization. When the metal is melted, the plates float on the metal forming a shingle-like cover preventing the products of combustion (Burner) from coming in contact with the metal surface. They are skimmed back to pour and used over and over. Graphite crucibles are made of German fire clay and graphite and fired in a kiln like pottery. It can be purchased in many grades of fineness from air float to fairly coarse. Super fine collodial graphite is used as a coating for die cast dies and permanent molds (metal molds) by suspending it in water and spraying the hot die or cavity. The water evaporates leaving a fine coat on the mold surface. It prevents the casting from sticking in the cavity or soldering itself to the cavity. This coat usually lasts for several castings. I have found over the years that anything short of the finest grade (most costly) is foolish finance.

The higher percentage of pure carbon the better. High carbon that is free from grit and foreign material is what to look for. There are quite a few prepared graphite based washes and facings on the market containing various synthetic and natural binders, additives and vehicles. They can be purchased dry (you mix it to suit yourself) or pre-mixed. One of the best all around commercial products I have found for both a mold and core wash is sold under the trade name of Zirc-O-Graph®A. It is a paste-like form of graphite and zircon flour with a binder which you mix with isopropol alcohol to the desired baume—a measure of specific gravity of liquids and solutions reduced to a simple scale of numbers.

Graphite can be produced artificially by passing an alternating current through a mixture of petroleum coke and coal tar pitch.

Each foundry supply manufacturer markets various types of washes and facings designed for molds, cores, ladle liners, etc.

graphite stopper: The stopper attached to the stopper rod of a bottom pour ladle (Bessemer Ladle). A stopper attached to a rod used in a large reservoir used to pour a mold.

gravel: Gravel refers to any granular material. An appreciable portion would pass through a sieve with ½-inch openings but be retained on a No. 6 sieve (six openings per square inch).

green bond strength: This is the strength of a tempered sand expressed by its ability to hold a mold in shape. Sand molds are subjected to compressive, tensile, shearing and transverse stresses. Which of these stresses is more important to the sands' molding properties is a point of controversy.

This green compressive strength test is the most used test in the foundry.

A rammed specimen of tempered molding sand is produced that is 2 inches in diameter and 2 inches in height. The rammed sample is then subjected to a load which is gradually increased until the sample breaks. The point where the sample breaks is taken as the green compression strength.

green compression: The measure of compressive strength of a green sand sample—tenacity.

green permeability: The measure of the ability to pass gasses through a sample of tempered molding sand.

green sand molds: A mold composed of prepared molding sand in the moist or as mixed condition.

grid bars and grids: Large flasks where the drag is rolled over and a bottom board would be cumbersome or impractical (Fig. G-8). The drag is fitted with cast iron or steel grids. The drag is rammed part way, the grid bolted in place and finished through the openings in the grid.

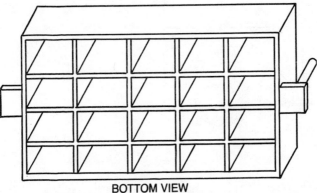

BOTTOM VIEW
GRIDDED DRAG

Fig. G-8. Grid bars and grids.

gyratory electric riddle: This piece of equipment is inexpensive by comparison to all other mechanical sand conditioners available (Fig. G-9). The Comb Electric Riddles have been manufactured for many, many years. I have one that was built in 1925 and it is still going strong. That company is still in business and a brand new one can be bought today. The Comb Riddle is simply an electric riddle or sifter which by its gyratory actions tempers and aerates the sand in one operation. The term *riddle* means to sift through a screen.

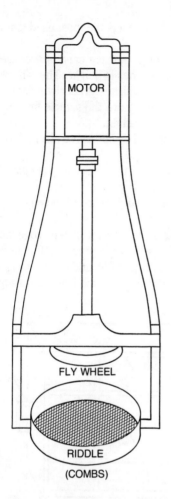

Fig. G-9. Gyratory electric riddle.

The operation of a Comb Riddle is simplicity itself. A one-sixth horsepower with a low speed motor at its top drives a shaft which is attached to a fly wheel directly above the riddle. This fly wheel is out of balance by virtue of a lead weight attached to one side of the flywheel's inner rim. When the fly wheel is rotated, it imparts a gyratory or shaking motion to the lower screen causing the sand shoveled into the riddle to mix and go through. The screens are easily removed by a quick clamp arrangement for cleaning or changing. The screens can be purchased in a wide variety of openings from 8 to the inch to the ½ inch. The ¼ mesh or four squares per inch is the most commonly used screen to condition sand.

hanging core: A core which is hung from the cope by tie wires (Fig. H-1).

hardener: Any metallic or nonmetallic element added to a base metal or alloy to increase its hardness. Antimony added to lead will increase its hardness.

hard spots: A problem with metallic iron castings. Wherever there is a shrink it will cause a hard spot, and as a result machining difficulties.

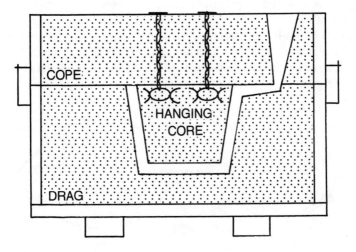

Fig. H-1. Hanging core.

hay rope: A loose rope made of hay used to wind around core arbors. When making large loam or dry sand molds the rope provides a cushion to aid in the core collapse and ample venting.

head strainer: A strainer core (ceramic or core sand) placed in a riser or head when the casting is poured directly into the head (Fig. H-2).

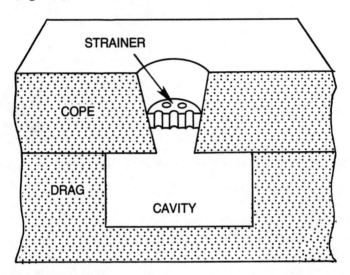

Fig. H-2. Head strainer.

heap sand: The sand used in the foundry to produce the bulk of the work, also called system or floor sand. Exclusive of special facing sands.

heart trowel: The heart trowel is a handly little trowel for general molding and as its name implies has a heart-shaped blade. They run in size from 2 inches wide to 3 inches wide in ¼ inch steps (Fig. H-3).

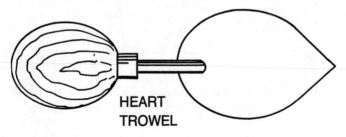

Fig. H-3. Heart trowel.

heat treating: The physical properties of some metals such as hardness, ductility and strength may be altered by various methods of heating and cooling.

high lead tin bronze: These alloys are traditionally the bearing and bushing bronzes. A typical high lead bronze would be: 80 percent copper; 10 percent tin; 10 percent lead; .32 pounds, per cubic inch; patternmakers' shrinkage ⅛ inches per foot; pouring temperature light castings 2,000 to 2250°F; heavy castings, 1850 to 2100°F.

high strength yellow brass: Another name for manganese bronze. Typical Composition: 63 percent copper; 25 percent zinc; 3 percent iron; 6 percent aluminum; and 3 percent manganese.

hinge tubes: Tin forms used to core holes through ribs and lugs. They extend into the mold perpendicular to the parting (Fig. H-4). The hinge tube is placed in a properly fitted notch in the

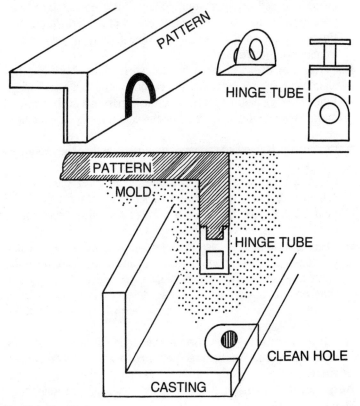

Fig. H-4. Hinge tubes.

pattern before the sand is rammed and remains in the mold after the pattern is drawn. Clean holes accurate for size and location are assured. Hinge tubes come in sizes from ⅛ inch hole to 1 inch and oval from 3/16 × 9/32 to ½ inches × ¾. They come ready to use with the pin tube filled with baked core sand.

holding furnace: An electric or fuel fired furnace for maintaining molten metal from a larger furnace at the right casting temperature. The holding furnace is not a melting furnace and is able only to maintain the temperature of a metal bath. It is used to collect and hold.

hot box process: The hot box process of making cores consists of accelerating the curing of a furan type binder with heat. This is done by ovens, infrared lamps, dielectric ovens or by heating the core box itself.

The application of heat greatly enhances the resin reaction. Originally Furan binders were the same used in the no-bake system (cured at room temperature). However, phenol has greatly replaced or substituted all or part of the furfural alcohol. Pnenol urea-formaldehyde or modified furan, phenol, furfural alcohol urea-formaldehyde.

hot chamber die casting: A die casting machine where the shot chamber is immersed under the casting metal and is self-filling. When the shot piston on its return stroke passes a port, the liquid metal then fills the shot chamber through this port for the next shot.

hot spruing: Removing castings from gates before the metal has completely solidified. Hot spruing of light section casting and intricate castings is widely practiced. The gates must be designed so that they will not break back into the casting. Small and medium gray iron casings can be cold sprued in most cases, leaving only a nub to grind.

hot tears: This defect is actually a tear or separation fracture due to the physical restriction of the mold and or the core upon the shrinking casting. The biggest cause is too high a hot strength of the core or molding sand. These defects can be external or internal. A core that is overly reinforced with rods or an arbor will not collapse.

If you restrict the movement of the casting during its shrinking from solidification to room temperature, it will literally tear itself apart.

humectant: Any material added to foundry sand to retain moisture.

DEGREES BAUME

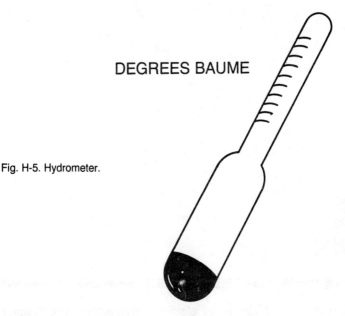

Fig. H-5. Hydrometer.

hydroblast: Cleaning castings with a high pressure stream of water and abrasive (using sand).

hydrogen: Hydrogen is a colorless, odorless, tasteless elementary gas. It is easily produced by the diassociation of water or from any hydrides.

hydrogen absorption: The absorption of hydrogen by molten metal. The hydrogen usually comes from the products of combustion cracking hydrides as water vapor H^2O. Some metals have a great affinity for hydrogen when heated, aluminum in particular. Care must be exercised when melting with gaseous fuels so that the flame is properly controlled. With aluminum it should be neutral or slightly oxidizing.

hydrometer: An instrument used to check the *baume*, the specific gravity of liquids (Fig. H-5).

hypereutectoid steel: Steel containing more than the eutectoid percentage of carbon.

I

infrared heat: The use of infrared heat to bake core with water soluble binders and those that soften on heating and harden on cooling (such as rosin).

illites: A type of bonding clay. Not all illite clays are suitable but the ones with high bonding power have good dispersion and are quite satisfactory.

impact test: The test given to a material with a pendulum in which the specimen is supported at one end as a cantilever beam (Fig. I-1). Measures the energy required to break off the free end. Called *Izod test*.

impoverishment: The loss of any constituent from an alloy or localized area of an alloy by oxidation, liquidation, volatilization or changes in the solid state. Also called *depletion*.

inclusions: Dirt, slag, etc. This defect is caused by failure to maintain a choke when pouring, dirty molding, failure to blow out mold properly prior to closing, sloppy core setting causing edges of the print in the mold to break away and fall into the mold. The drag should be blown out, the cores set and blown out again. Dirt falling down the sprue prior to the mold being poured or knocked in during the weighting and jacketing.

inconel: A high nickel metal used for furnace fixtures, dairy and food equipment. 79.5 percent nickel, 13 percent chromium, 6.5 percent iron, 0.08 carbon, developed by International Nickel Co.

ingates: The gate leading directly into the mold cavity.

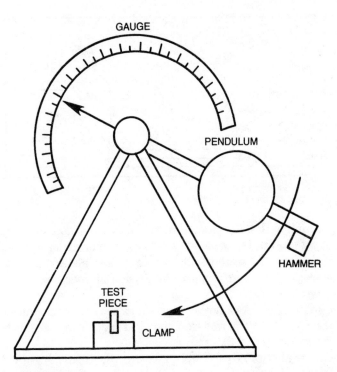

Fig. I-1. Impact test.

inhibitor: A material such as fluoride, boric acid or sulphur added to the molding sand for casting magnesium alloys to prevent burning of the molten magnesium. Restraining an undesirable chemical action. Any addition to a solution or substance to prevent or minimize corrosion.

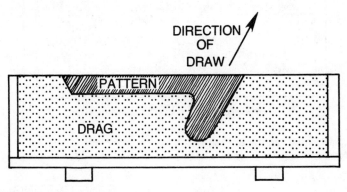

Fig. I-2. An irregular draw.

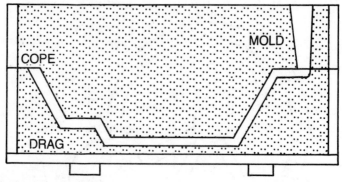

Fig. I-3. Irregular parting.

ingot molds: Cast iron or steel molds for casting ingots also used to pig or ingot the metal left over in a ladle or the days melt.

innoculant: Any material added to a molten metal which will modify the structure changing the physical and mechanical properties.

irregular draw: A pattern or portion of a pattern so constructed that it must be removed from the mold in other than a straight vertical lift (Fig. I-2).

irregular parting: A parting plane between the cope and drag that lies on more than one plane (Fig. I-3).

Inserts: Parts formed of a second material, usually a metal. They are placed in a mold and cast into the body of the casting and appear as integral structural parts of the final casting (Fig. I-4).

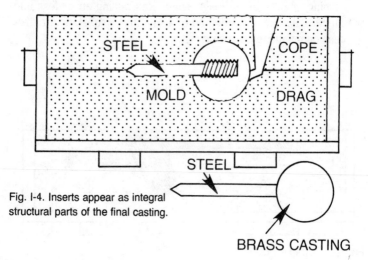

Fig. I-4. Inserts appear as integral structural parts of the final casting.

inverse chill: A defect is found in gray iron castings. It is hard or chilled iron. The cause again could be incorrect carbon equivalent for the job or the presence of nonferrous metals in the charge, lead, antimony or tellurium. They are detrimental impurities.

investment casting wax: Any wax used for expendable wax patterns and gates. It runs from a hard carving wax to extremely soft. Over 2,000 formulas are available in natural or synthetic waxes.

investment molds: Molds constructed of a refractory material around wax or other expendable patterns such as plastic, mercury, etc.

J

jacket: A jacket is a wood or metal frame which is placed around a mold made in a snap flask during pouring to support the mold and to prevent a run out between the cope and drag.

The common practice is to have as many jackets (For each size snap mold) as can be poured at one time. When the molds have solidified sufficiently the jackets are removed and placed on the next set of molds to be poured. It is called *jumping jackets*. Jackets must be carefully placed on the molds so as not to shift the cope and drag. The jackets must be kept in good shape and fit the molds like a glove. A tapered snap flask and tapered jackets work best due to the taper.

jack stars: Jack stars are white cast iron stars similar to the jacks children play, except that they are pointed (Fig. J-1). These starts come in various numbers of points and sizes and are used to clean castings in a tumbler. The castings and stars are charged into the tumble and the tumbler is rotated. The rotation causes the stars to rub and clean the castings. The points of the stars pick and rub sand loose from the castings. They can be purchased, but most iron foundries have a matchplate of star patterns and cast them as needed.

jar ramming: Also called *jolt ramming*. The packing or ramming of the sand in a mold or core box by raising and dropping upon an anvil, the sand itself being the ramming medium (Fig. J-2). Jolt machines come in various sizes from small bench jolt machines to

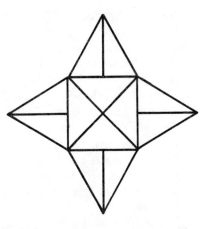

Fig. J-1. A white iron jack star.

large machines which will handle an extremely large burden (weight). These machines are pneumatic in operation. The air enters the cylinder under the piston raising the table and mold when the piston passes the exhaust port. The air behind the piston exhaust and the table and mold fall striking the anvil.

The abrupt stopping of the table, mold and sand upon striking the anvil does the ramming as the sand that is free to move continues its trip downward. The cycle is repeated until the desired density of the ram is reached.

Jolt machines can be purchased with various features which roll the mold over and draw the pattern when the jolting is finished.

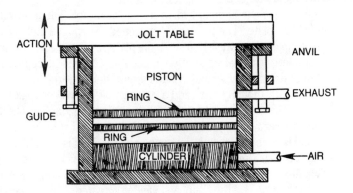

Fig. J-2. The process of jar ramming.

jet engine bronze: An alloy of 68 to 62 percent nickel, 21 to 24 percent molybdenum, 7 to 8.2 percent aluminum (less than .1 percent carbon, .25 percent silicon, .15 percent iron and .30

percent manganese). This alloy is called *kinsalloy* and is used for high-temperature jet engine parts.

job shop: A foundry which engages in the casting for all comers as opposed to the captive shop that manufactures numerous types of castings in small quantities.

jolt pin lift machines: Operated by the same principal as the stripper. The foundrymen lift the rammed flask from the pattern. The only difference is that it is done by four pins engaging the four corners of the flask. The pins are carried on a yoke which is lifted by suitable cylinders.

jolt squeeze pin lift machine: A jolt squeeze machine equipped with a pin lift to draw the mold from the pattern which is fixed to the table.

Unlike the jolt squeeze machine on which the entire mold is produced (cope and drag), the jolt squeeze pin lift, jolt stripper, jolt rollover, jolt rollover and draw and the jolt pin lift machines produce half molds. The molders work in pairs. One operator makes drags and one makes the matching copes.

The usual combination is a jolt rollover draw or jolt squeeze rollover, a draw machine for the drag (which must be rolled over) and a jolt pin lift or jolt squeeze pin lift for the cope.

Completely mechanized machines can be had to make the mold completely ready to pour.

jolt squeeze molding machine: A molding machine which is equipped with two cylinders and two positions—one for jolting the mold and one for squeezing the mold (Fig. J-3).

The machine is equipped with a molding table and a movable squeeze plate. The machine is designed to produce molds from matchplates. In operation, with the platen swung aside, the flask complete with pattern between cope and drag is placed on the table, drag side up. Sand is riddled onto the pattern, the drag is filled with molding sand and a bottom board is placed inside the drag. The mold is jolted (see *jar ramming*) by the molder operating the jar air valve with his knee. When jolted sufficiently, the mold is rolled over with the cope upward. Sand is riddled on the patterns in the cope, the cope filled and peen rammed and a squeezer board is placed inside the cope. The platen is swung into position and the squeeze cylinder actuated. This procedure raises the mold against the platen and squeezes the entire mold between the bottom board and squeezer board finishing the ramming of the cope and drag. The platen is swung aside the squeezer board, removed and the sprue is cut. The pattern is vibrated, the cope drawn and set off. The pattern (matchplate) is

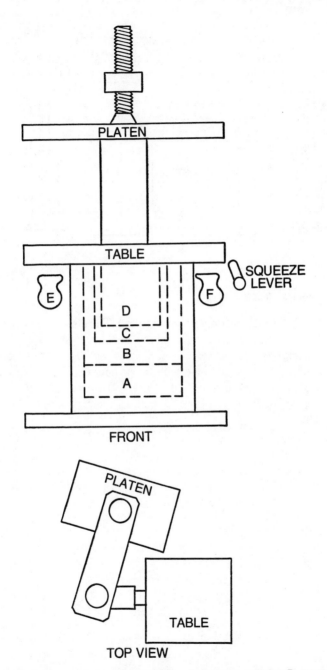

Fig. J-3. A jolt squeeze molding machine. A. squeeze cylinder; B. squeeze piston; C. jolt piston; E. vibrator knee valve; and F. jolt knee valve.

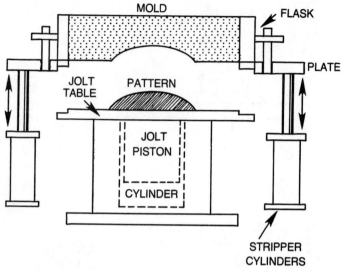

Fig. J-4. Jolt stripper machine.

again vibrated and drawn from the drag. The cope is then placed
on the drag and the completed mold is set off for pouring.

jolt stripper machine: See *jar ramming*. The jolt stripper
machine is a jolt machine with an air-operated lifting plate on
which the flask bearing sets (Fig. J-4). This plate is raised
drawing the mold half from the pattern which is fixed to the jolt
table.

K

kaolinite: A hydrated aluminum-silicate clay also known as *fire clay*. It is one of a family of three distinct minerals having a similar composition—kaolinite, nacrite and dicite. It is a sedimentary clay of low flux content. Its prime use in molding sands is in facing for very heavy work and also in dry sand work when excellent hot strength is required. It is often blended with western bentonite for this purpose. It comes closer than any other type of bonding clay to producing a sand which approximates the properties of naturally bonded sands. It's tough, durable and easy to use.

killed steel: Steel that has been completely deoxidized with silicon or aluminum to reduce the oxygen content to a minimum.

kiln dried: The air drying of pattern lumber artificially in place of drying naturally.

kish (cast iron): If the carbon equivalent of the iron is too high for the section poured and its cooling rate is slow, free graphite will form on the cope surface in black shiny flakes free from the casting. It causes rough, holey defects usually widespread. The carbon equivalent is the relationship of the total carbon, to the silicon and phosphorous content of the iron. It is controlled by the make-up charge. Silicon is added at the spout. It is called carbon equivalent because the addition of silicon or phosphorous is only one third as effective as carbon, therefore the carbon equivalent of the three additives is equal to the total percentage of pure carbon plus one third the percentage of the silicon and phosphor-

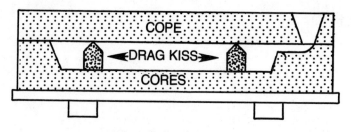

Fig. K-1. A kiss core does not set in a print.

ous combined. The carbon equivalent is varied by the foundryman depending on what type of iron he is producing and what he
is pouring.

kiss core: Kiss core is a small core which does not set in a print, but
is held in place by the pressure of the cope surface of the mold
(Fig. K-1).

knock out: The operation of removing sand cores from a casting.

L

ladle additions: See Innoculant.

ladle bowls: Hand shank bowls which are shaped more like a bowl than a bucket (Fig. L-1).

ladle shanks: The device by which a ladle is carried for transporting and pouring.

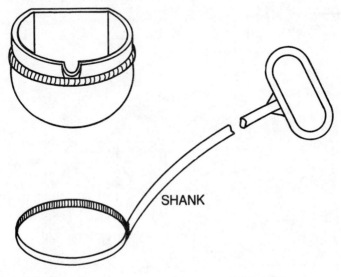

SHANK

Fig. L-1. Ladle bowl.

ladles: Steel, refractory lined containers used to receive and pour molten metal. They range in size from small hand ladles to crane ladles of 100 tons and over in capacity.

lagging of patterns: Increasing the size and dimension of a pattern by applying strips of wood, plastic or sheet wax to the area you wish to increase in dimension.

lake sand: Another name for river sand, sharp sand, bank sand, white sand, core sand and sugar sand. Usually defined as clay-free high silica sand.

lamp black: A practically pure carbon (soot) produced by burning carbonaceous substances in an insufficient supply of air (reducing atmosphere). It is used in core and mold washes and blackings.

large skim gates: Skimmer tins (Fig. L-2). Round and perforated tins or mica used as a skimmer under the pouring basin or elsewhere in the gating system.

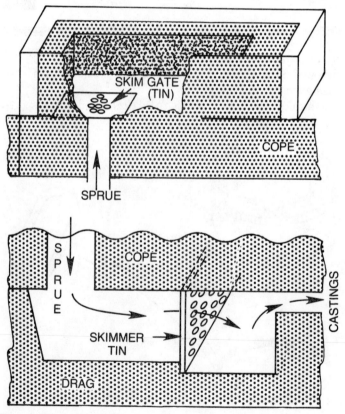

Fig. L-2. Large skin gates.

latent heat of fusion: When a molten mass of pure metal starts to cool the temperature drops at a definite ratio. A point of solidification is reached where the temperature remains constant until the entire mass is solidified. This point is called the *latent heat of fusion*.

lathe dog: A forged or cast L-shaped device used to drive work in the lathe when turning between centers (Fig. L-3). The work is clamped in the dog jaw with a set screw or cap screws. The drive leg fits loosely into a slot in the faceplate.

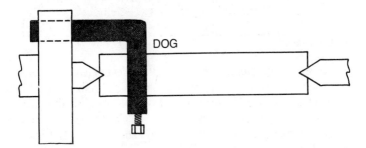

Fig. L-3. Lathe dog.

launder: An old term used to designate a channel for conducting molten metal.

layout board: A smooth board with two sides square upon which the pattern layout is made. The dimensions are taken from this layout when making the pattern. A blueprint is used only to make the layout as they never show core prints, core boxes, etc., which the patternmaker must determine himself and put on the layout board.

lead: A soft heavy bluish metal, Pb, obtained chiefly from the mineral Galena. Melting point 621°F, boiling point 2777°F. When remelted from scrap it is called secondary lead. It is widely used in alloying with other metals.

leaded cores: A cousin of bolted cores (Fig. L-4). In place of bolting a core together you can lead them together by pouring molten lead into the attachment holes and then covering the top counter sink with a cap core. The cores must fit very closely to prevent lead from running out at the joint.

lead dispersement: The amount of lead that will go into solution with another metal. For example, pure molten copper will dissolve approximately 35 percent lead. If 2½ percent nickel is added, it jumps up to 70 percent.

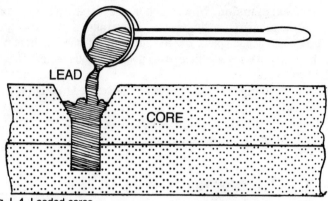

Fig. L-4. Leaded cores.

lead effect: The reference here is to the effect of lead as a defect.

The presence of lead in a heat of aluminum bronze can cause a defect called *blue*. These are blue-to-purple stains on the surface of the casting, usually from cores or chills as little as .005 to .01 percent of lead amounts. Over .03 will cause gray to black stains.

It has little effect in regard to the physical other than to contribute to hot shortness at elevated temperatures. Lead in the amount of .40 percent and above in manganese bronze (high strength yellow brass) has a deleterious effect on the mechanical properties and should be kept to .20 percent or lower. Lead in silicon bronze combines with the silicon to form lead silicate (glass). The problem is in distinguishing some red brasses (leaded) from silicon bronze. A small chunk of a leaded red brass in a silicon bronze heat will spoil the entire heat.

leaded nickel brass: An alloy of copper, tin, lead, nickel, zinc and iron. Also called nickel silver. When the alloy contains from 11 to 17.5 percent nickel in its composition it is called *leaded nickel brass*. When the nickel content is from 19.5 percent to 27 percent, it is called *leaded nickel bronze*.

Leaded Nickel Brass—Typical:

- Copper, 58 percent
- Tin, 3 percent
- Lead, 11 percent
- Nickel, 14 percent
- Iron, 1.5 percent
- Zinc, Balance

Leaded Nickel Bronze—Typical:

- Copper 67 percent

- Tin 4.5 percent
- Lead 5 percent
- Nickel 21.5 percent
- Zinc 10 percent
- Iron 1 percent
- Manganese 1 percent

As the nickel increases the melting and pouring temperature increases. Percent nickel for 14 percent nickel, the pouring temperature for medium size castings would be 2200°F. For 17.5 percent nickel, 2260°F; 21.5 percent nickel, 2300°F; and for 24 percent nickel, 2400°F.

leaded red brass: The most common leaded red brass is called ounce metal or 85 three 5's: 85 percent copper, 5 percent tin, 5 percent lead and 5 percent zinc. It is .318 pounds per cubic inch and a 3/16 patternmaker's shrinkage per foot. The pouring temperature for light work is 2100°F to 2300°F, for heavy work, 1950°F to 2150°F. It can be deoxidized with 15 percent phosphorous copper, 1 to 2 ounces per 100 pounds.

An easy alloy to cast for general work. When the tin content is below 4 percent and the zinc increases it is called *semi-red brass*.

Semi-Red Brass—Typical:

- 78 percent copper
- 3 percent tin
- 7 percent lead
- 12 percent zinc

leakers: Castings when subjected to hydraulic pressure leak through the casting walls. This can be caused by many factors such as pouring temperature, gas porosity, wet molds, design, shrinkage porosity or a chaplet which did not fuse or burn in properly. In short, it is an unsound casting. More often than not the fault lies in the casting design.

lead sweat: High lead tin bronzes containing 15 to 27 percent lead are inclined to bleed or sweat lead. Also they have lead segregation. Excess gas absorption during melting will accelerate this condition. It cannot always be cured with excessive phosphorous copper additions (to deoxidize). It helps to add a small amount of lead oxide or copper oxide to melt 3 to 5 minutes before pulling the pot.

Remember that these alloys start to melt at 620°F, the melting point of lead. If you shake the casting out of the mold before it gets down below 620°F, it is going to bleed all the lead out from between the copper crystals and leave you with a lead-coated copper sponge. It is best to leave these high lead castings in the sand until they are at room temperature.

leaves in core: Strange as this may seem, blowing and gas porosity in a casting are often traced to leaves, roots, coke and coal in the core sand. When using a raw beach or lake sand, if you do not buy washed and dried silica sand, check your sand for organic matter. It might be so very finely divided that it can escape detection. Check using the loss to ignition test to determine how much trash is present in the sand other than silica.

lifting force: The lifting or buoyant force on the cope is the product of the horizontal area of the cavity in the cope, the height of the head of metal above this area and the density of the metal. This lifting force on the cope has to be reckoned with. If the force is greater than the weight of the cope, the cope will be lifted by this force when the mold is poured and run out at the joint resulting in a lost casting plus a mess. In order to determine how much weight has to be placed on the cope, if any, to prevent a lift of the cope we mulitply the area of the surface of the metal pressing against the cope by the depth of the cope above the casting and the product by 0.26 for iron, 0.30 for brass and 0.09 for aluminum. These figures represent the weight in pounds of 1 cubic inch of iron, brass and aluminum respectively.

So if you are casting a flat plate 12 × 12 inches, the surface area against the cope would be 144 square inches. If this was molded in a flask which measured 18 × 18 inches with a 5-inch cope and the metal you are casting is red brass, you have 144 × 5 inches the height of the cope, times 0.30 (the weight of 1 cubic inch of brass). You wind up with a lifting force of 216 pounds. One cubic inch of rammed molding sand weighs 0.06 pounds. So with that factor, the cope weight would be 18 × 18 × 5 inches times 0.06 or 97.2 pounds. If you subtract the cope weight from the lifting force of 216 pounds minus 97.2 pounds, you are still short by about 119 pounds. This would be the weight you would have to add to the cope to prevent it from lifting with no safety factor. With a 20 percent safety factor, add 24 more pounds. You should be put 150 pounds on the cope.

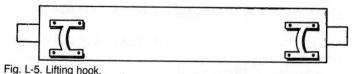

Fig. L-5. Lifting hook.

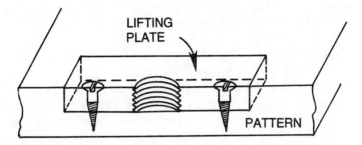

Fig. L-6. Lifting plate.

lifting hook (flasks): See also *grab hook*. A hook or pad eye attached to a large flask for lifting or rolling over (Fig. L-5). A flask trunnion is sometimes called an *open lifting hook*.

lifting plates: Metal plates let into a pattern with a tapped hole in which a lifting screw (eye) with mating thread is screwed into in order to lift the pattern (Fig. L-6).

light metals: Metals that have a low specific gravity such as beryllium, magnesium and aluminum.

light metal gate tubes: Perforated tin tubes used as skimmers in pouring aluminum and magnesuim—strainer tubes (Fig. L-7).

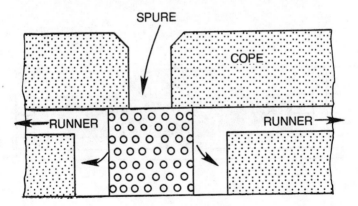

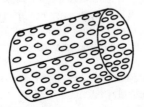

Fig. L-7. Light metal gate tubes.

lime in molding sand: The presence of excessive lime in a molding sand is often the hidden cause of gas holes, porosity, pin holes and blisters in a casting.

liquid contraction: The shrinkage occurring in metal in the liquid state as it cools. Volumetric shrinkage.

liquid parting: A liquid used as a parting material in place of a dry or dust type parting, brushed or sprayed on the pattern. Old time liquid parting was bayberry wax dissolved in gasoline. When the gasoline dissolved the pattern was lightly coated with wax.

There are numerous commercial liquid partings on the market, ground mica, graphite, waxes etc. They are dissolved or dispersed in various vehicles.

liquidus: The temperature at which freezing begins during cooling, or melting ends during heating.

litharge: Lead monoxide, PbO, also called *massicot*. It is yellow-to-orange powder used in the foundry to make oil sand followers (oil match) (Fig. L-8). The most used formula is ½ dry naturally bonded molding sand and ½ dry sharp (core sand). Temper the sand with linseed oil and mix in a small handful of litharge.

When dry, a litharge match or follower is rock hard and will take a lot of abuse lasting for years.

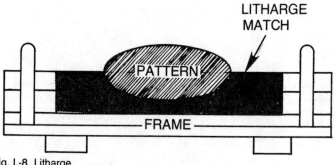

Fig. L-8. Litharge.

loam molding: The production of large molds built up with bricks, plates and various arbors which are covered with loam and dried to give the form of the casting desired. A high skill is required, covering a large variety of foundry knowledge, molding, pattern making, melting, etc.

loam sand: A mixture of sand, silt and clay in such proportions as to exhibit sandy and clay particles in about equal proportions: 50 percent sand and 50 percent clay and silt.

lock buttons: Aluminum buttons riveted to a matchplate directly over a corresponding counterbore recess (Fig. L-9). When the mold is made and the plate removed, you have sand projections on one face of the mold and corresponding female depressions directly opposite on the other face. When the mold is closed, these male and female sand members provide a sand to sand lock to help prevent the mold halves from shifting during handling, jacketing, pouring, etc.

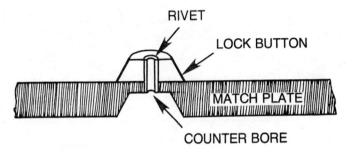

Fig. L-9. Lock button.

loose patterns: Unmounted patterns.

loose piece: A part of a pattern so attached that it remains in the mold and is removed after the main body of the pattern is drawn (Fig. L-10). The loose piece might be dovetailed to the pattern or pinned on with dowels. If by dowels, the sand is packed (rammed) tightly around the dowels and the loose piece to hold it in postion. The dowels are withdrawn and the mold finished in the usual manner.

loss and gain in the cupola: The loss or gain of various elements when melting with a cupola will vary depending upon its operation or who is operating it, however, the average when melting iron is loss of silicon, 10 percent; loss of manganese, 20 percent; and a gain of sulphur, .03 to .05.

When melting brass; the loss of copper, nil; loss of tin, negligible; loss of lead, 15 to 25 percent depending upon the percent of lead in the alloy. The higher the percent the greater the percent loss and loss of zinc, 50 to 60 percent.

When melting brass or bronze in the cupola, it is best to melt the copper and tap it into a hot ladle containing the tin, lead and zinc in the form of shot or small preheated chunks.

loss of ignition: The test to determine the percent of material in a molding sand which is lost or burned away during casting.

Five grams of the sample sand are placed in a porcelain cruci-

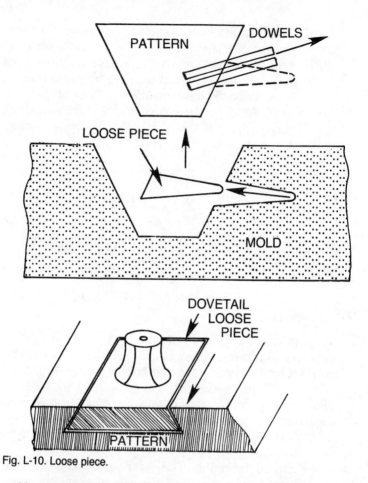

Fig. L-10. Loose piece.

ble and heated at 1610°F for one hour and reweighted:

$$\frac{\text{Loss in weight}}{\text{Weight of Sample}} \times 100 = \text{percent of Loss}$$

Too high a percent of carbonous or gas-producing materials in a molding sand will cause gas holes, porosity and pin holes in the castings. The percent is checked by ignition test.

lug: An ear-like projection, frequently split as a clamping lug on a tailstock.

lugs, matchplate: Metal lugs attached to the ears of a matchplate that match the flask pin shape (Fig. L-11). The lugs follow the drag pins as the plate is lifted (drawn) guiding the plate straight up.

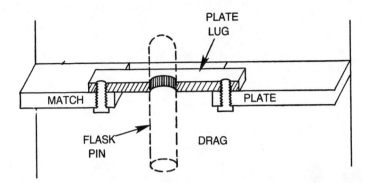

Fig. L-11. Matchplate lug.

lumber, diamond pattern: Pattern lumber with a carved face in various diamond designs. This lumber is used to make patterns with a non-skid surface such as treads, sewer covers, catch basins and similar industrial and street castings. Five different diamond designs are available from most patternmaker's supply houses, in thicknesses of ½ inch, ⅝ inch and ¾ inch thick, 8 inches, 10 inches and 12 inches wide and in various lengths. The design is machine carved and clean with ample draft.

lute: A mixture of fire clay used to seal cracks such as the crack between a crucible and a crucible cover.

lycopodium: A fine yellow powder obtained from a plant known as club moss, cultivated in Russia. It is regarded as the finest and perfect dry parting material, used when casting highly ornamental bronze work (as a dry parting). It is expensive, therefore, is used only for that special job. It can be purchased in some drugstores.

M

machinability: A measure of the ease of machining a metal. The easiest material to machine and that which offers the least resistance to machining would be rated 100. As a material becomes tougher to machine the rating lowers. Where a leaded red brass would have a machinability rating of 90, an alloy of 90 percent copper and 10 percent nickel would have a machinability rating of 10 or less, white iron 0 to 1.

machine ramming: The term used to refer to a mold on a molding machine, slinger, etc., other than hand ramming. The use of pneumatic rammers is considered hand molding.

magnesium: A silver-white metal Mg, 64 percent the weight of aluminum with the specific gravity of 1.74. The melting point of magnesium is 651°C, 1202°F and the boiling point 1120°C (Fig. M-1).

Magnesium and magnesium alloys are melted and poured from crucibles made of welded boiler plate in sizes from 60 to 1000 pounds capacity. Small shops melt in the ladle. The ladle simply sits down through a hole in the furnace cover.

Sand used for magnesium casting must be low in moisture and contain inhibitors to prevent oxidization of the magnesium from water vapor. See inhibitors.

Multiple gates should be used in place of one or two large gates in order to fill the mold as rapidly as possible to prevent the sand from getting too hot and causing excessive steam. A typical magnesium base alloy for sand casting is composed of aluminum,

9 to 11 percent; zinc, .3 percent maximum; silicon, .3 percent maximum; copper, .10 percent maximum; nickel, .01 maximum; and magnesium, the balance. The tensile strength as cast is 23,000 PSI, when heat treated, it is 32,000 PSI.

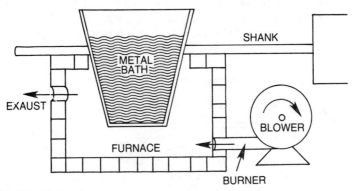

Fig. M-1. Magnesium.

magnetic brad set: A pattern and model maker's tool used to set (drive) small wire brads with ease and in hard-to-reach places (Fig. M-2). It is a wooden handled tool equipped with a hollow tube which is spring loaded into the handle. The brad is held in the tube by the magnetic tipped plunger. The nose of the tube containing the brad is placed where you desire to set the brad. Force is applied to the handle pushing the magnetic plunger forward as the tube goes back against its spring thus pushing the brad into place.

The brad is pressed in place with a steady firm pressure. Never strike or drive the tool. This results in a bent brad and marred surface.

magnetic pulley: The magnetic pulley is a pulley used on a sand conveyor belt in ferrous foundries to remove tramp metal such

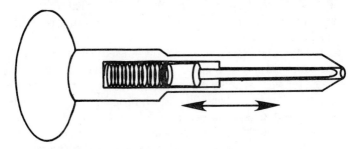

Fig. M-2. Magnetic brad set.

as lost gates and shot metal from the sand (Fig. M-3). Since it is a permanent magnet, the sand will discharge from the belt and the tramp iron or steel will continue part way around the pulley due to magnetic attraction. It will discharge when the belt carries it away from the magnetic bond or attraction, dropping it off under the belt. Eriez of Canada produces a magnetic pulley with a unique magnetic field which is criss-crossing, radial, axial and diagonal. It eliminates dead spots.

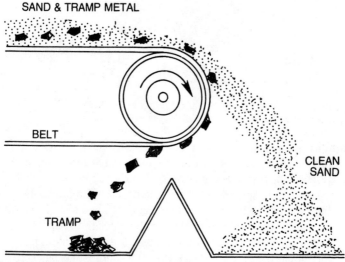

SAND & TRAMP METAL

BELT

CLEAN SAND

TRAMP

Fig. M-3. Magnetic pulley.

manganese copper: An alloy of manganese and copper used for the introducton of manganese into an alloy.

marine animal oils: a class of marine oils such as menhaden oil, sardine oil, whale oil, etc. These oils are used in conjunction with a vegetable oil such as linseed oil to produce core oils. (Menhaden oil is made from boiling porgy fish. Also called white fish).

mass hardness: A condition in which the entire casting is too hard and unmachinable.

master dowels: Male and female dowels used in pattern to insure a correct and precise register of pattern components while at the same time allowing the components to be freely taken apart or put together for molding (Fig. M-4). Maintains correct register with repeated use, which is not the case with wooden dowels.

Master dowels can be purchased in a large variety of sizes, designs and lengths in brass or steel with the special tools required to install them.

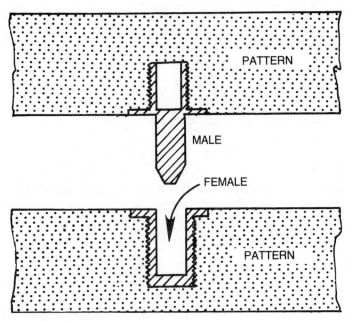

Fig. M-4. Master dowels.

master pattern: A pattern used to cast the production patterns or equipment.

master sheet wax: A sheet wax which can be purchased in a wide variety of thickness. It can be purchased with or without pressure sensitive adhesive on back. Its uses are too numerous to put down. The limit depends upon the ingenuity of the user. The wax can be purchased with melting temperatures from 100°F to 300°F in sizes from 8 inches × 12 sheets to 12 inches × 24 inches sheets with a thickness of from 1/64 inches to ⅜ inches. Its use in the pattern shop is to *lag* up a pattern, make a core box pattern from a *core plug*, drier patterns from plugs or a master pattern from a plug.

In the event only a few castings are required, a plug can be lagged up with sheet wax to the desired metal thickness and shape, shellacked and used as a pattern.

matched partings: Forming of a projection upon the pating surface of the cope half of a pattern and a corresponding depression in the surface of the drag.

matchplate: The matchplate is the same as the mounted pattern with the exception that when you have part of the casting in the cope and part in the drag (split pattern), these parts are attached

to the board or plate opposite each other and in the correct
location so that when the plate is removed and the mold is closed,
the cavities in the cope and drag match up correctly (Fig. M-5).

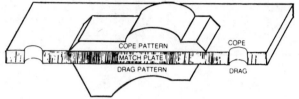

Fig. M-5. Matchplate.

matchplate flask: A large steel flask (very accurate) fitted with
lifting screws to make a perfect draw used to cast matchplates
(Fig. M-6).

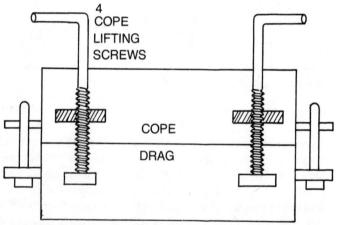

Fig. M-6. Matchplate flask.

matchplate inserts: Steel buttons cast in aluminum matchplates.
When they are produced, these buttons are located where the
cope and drag flask rest on the plate during molding (Fig M-7).
Their purpose is to prevent the flask from wearing the relatively
soft aluminum plate at its point of contact.

Fig. M-7. Matchplate inserts.

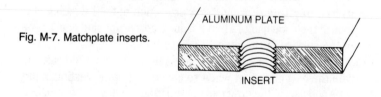

medium pattern: A pattern that is used only occasionally or for casting a one time piece. It is usually constructed as cheaply as possible. If it is a split pattern, wooden dowels are used for pins and fit it into holes drilled into the matching half.

melamine: A trimmer of cyanide used to produce a resin by reacting melamine and formaldehyde. A core binder.

melting atmosphere: The atmosphere in the melting furnace which is reducing, (short on oxygen), oxidizing (rich on oxygen), neutral (neither oxidizing nor reducing) or inert by the introduction of an insert gas into the melting chamber.

melting loss: The loss of metal in the charge during the melting operation which is kept to a minimum with good sound melting practice.

melting rate: The weight of metal in pounds or tons melted in one hour. If a cupola has a rate of 5, it should melt 5 tons of iron per hour at the tap.

melting ratio: The proportion of the weight of metal to the weight of fuel in a cupola melting. A cupola melting at a 7 to 1 ratio is melting 7 pounds of iron for each pound of coke.

metallic filler: Iron, aluminum, copper or brass powder mixed with an epoxy or resin used to repair minor defects in a casting face or surface.

metal penetration: This defect should not be confused with an expansion scab which is attached to the casting by a thin vein of metal. The defect is a rough unsightly mixture of sand and metal caused by the metal penetrating into the mold wall or the surfaces of a core (not fused).

 The defect is basically caused by a too soft and uneven ramming of the mold or core, making the sand too (open) porous. A too high pouring temperature or a corner too sharp (insufficient filleting) makes it impossible to ram the sand tight enough, localize the overheating of the sand due to poor gating practice or mold with a sand that is too open for the job.

 Penetration in brass castings is sometimes traced to excessive phosphorous used in deoxidizing the metal which makes it excessively fluid.

misrun: A portion of the casting fails to run due to cold metal, slow pouring, insufficient hydrostatic pressure or sluggish metal (nonfluid due to badly gassed or oxidized metal).

micro shrinkage: The reference to extremely fine micro porosity caused by shrinkage.

moisture: The percent of moisture (water) in a molding sand.

mold and core washes: These washes consist of coating the internal sand mold cavity and the surface of the core which comes in contact with the hot metal with a *refractory* coating in order to prevent the metal from burning-in. They produce smoother surfaces on the castings and facilitate the separation of the castings from the sand. This process is known as *peel*. In low melting metals and alloys this is not much of a problem due to the relatively low temperature involved.

Aluminum can be cast in molding sand which is quite fine and with a low refractiveness. It produces a clean smooth surface and excellent peel. However, as dry sand cores used in aluminum casting are usually quite open (high permeability), the cored cavity will not be nearly as smooth as the surface of the casting in contact with the green sand. And the usual practice is to coat the core with a *wash* to fill in the spaces between the sand grains and present a smooth surface of the metal to lay agains. Lead, zinc and very fluid metals which will actually seek their way between the grains of sand due to a low surface tension are examples of where you would use a mold or core wash.

As a general practice it is wise to coat the portion of all dry sand cores with a core wash. As the temperature of the metal being cast rises, the permeability of the sand must be increased to allow for the greater volume of gas and steam being generated to pass through the mold walls and not back through the molten metal (courser sand).

These conditions increase the need for mold and core washes.

Also, when the meal section is quite heavy and subjects the sand to a longer period of heating during solidification you need a wash. A ¼-inch brass casting poured at 2150°F would present no problem as to burning or peel, but the same metal poured even at a lower temperature of 1950°F and 1-inch to 2-inches thick in section would burn-in badly with poor peel. This also applies to pouring the metal at an excessive temperature.

Another cause for rough surface poor peel and burn-in would be a casting poured with an excessively tall sprue and risers above the highest point of the casting. Here we have increased the *hydrostatic pressure* greatly against the cope surface causing the metal to penetrate the cope giving a rough surface.

Where you have a casting which is quite large in size requiring a longer time to pour the section of sand directly below, the pouring sprue must take the impact and passage of hot metal for a longer interval of time and can be burned and eroded away washing into the casting. Here again is where a good wash would be used also at the gates into the cavity.

Where you have an internal core which is necked down and passes through a wall even though you are not concerned with the appearance or smoothness of the internal cavity, you have what is known as a hot spot. The core has to be washed at this point or the core would be badly eroded or come completely apart.

Learning when, where and how much to apply a mold and core wash is learned by experience and commonsense.

A mold facing or wash will overcome some of the problems of a poor sand that has insufficient refractories, but this is foolish finance.

Dry mold facings are also known as *shake-on facings*. Graphite is the most widely used.

Core and mold washes are carbonaceous and carbon free (Table M-1). Binders are organic and inorganic (Table M-2).

Table M-1. Carbonaceous and Carbon-Free Molds.

Carbonaceous	Carbon Free
Graphite	Silica Flour
Carbon Black	Talc
Ground Hard or Soft Coal	Magnesite
Petroleum Pitch	Alumina
	Asbestos
	Mica
	Zircon

In the proper selection of a mold or core wash, dry or wet, you are better off with a simple compound. *Skin dried* molds come under the same category.

There are two ways of accomplishing this treatment. One way is to spray the mold cavity with a wet binder with or without a carbonaceous or carbon-free ingredient.

On light-to-medium size castings usually the mold can be simply sprayed (fine spray) with a mixture of one part molasses or

Table M-2. Binders.

Organic	Inorganic
Starches	Clays
Dextrin	Bentonites
Sugar	Silicates
Molasses	Oxychloride
Rubber Cement	
Linglin Sulphite	
Glucose, etc.	

linglin sulphite to 10 parts of water. Then carefully dry this with a soft flame torch moving continuously to prevent localizing heat and burning. The mold will then have a dry firm skin upon which the metal can lie. The trick here is to dry without burning the binder (molasses or linglin sulphite) and the mold must be poured shortly after skin drying to prevent the moisture in the mold behind this skin from migrating back to the surface. If this migration occurred, it would defeat the purpose. For heavier work or higher melting metals you can add graphite, talc or any one of the bases and proceed just described.

A very good method when using green sand cores made for the same sand the mold is made from is to dry the core completely. Dry it in a core oven 350° to 400°F until dry through and through or overnight. When the cores are removed from the oven they are very fragile, crumbly and cannot be handled at all. If sprayed while hot with molasses, or linglin sulphite when cold, they have a nice firm skin and can be set in the mold. Here again you should pour the mold shortly after closing to prevent moisture in the mold from migrating to the core.

The cores must be cold. A hot core set in a cold mold will cause condensation and blow.

This system is not cheap but does a very excellent job. The cores shake out easily leaving a clean smooth defect-free surface. On thin wall castings, they can't be beat.

In place of skin drying the mold by spraying and torch drying, you can mix the binder with enough sand to cover the pattern with a layer ½ to 1-inch thick. Finish the mold with floor backing sand and then dry the mold face with a torch.

The binder can be molasses, linglin sulphite water, dry goulac, dry linglin sulphite, glucose, rosin, etc.

Now it is common practice to use a facing sand in many cases that is not skin dried but can consist of merely new sand, or sand mixed with sea coal, pitch flour, wheat flour, etc, and backed with flour or system sand.

A typical facing sand mix consists of three parts air float sea coal; one part Goulac; and 10 parts molding sand. Another mix consists of 15 parts sharp sand; one-half part bentonite; and three parts system sand.

In some classes of work the molder works with two sands, one for facing and a courser sand for backing. This can cause trouble as the two eventually become so mixed together in the system that it all becomes useless.

moldability: This characteristic is also related to the nature of the bonding clay and the fineness of the sand.

Because the base sand determines the resulting finish of the casting, it should be selected with care keeping in mind the type, weight and class of casting desired. The three or four types of screened sands formerly used for a base have given way to the practice of blending one coarse sand with a fine one which results in a better grain distribution. It has been found to produce a better finish and texture. Each of the two sands selected should have a good grain distribution within itself. Contrary to popular belief, additives of an organic or carbonaceous nature do not improve the finish but only furnish combustibles resulting in better peel.

mold clamps: Floor molds are always clamped to pour. Sometimes they are weighted, sometimes not, but always clamped (Fig. M-8). There are two basic types of clamps used.

The *C clamp* is cast iron or square steel stock. In operation the bottom foot is placed under the bottom board and a wedge (wood or steel) is tapped between the top foot and the top of the flask side.

A better practice is to place two wedges between the clamp foot and flask by hand, then tighten them with a small pinch bar. Driving them tight jolts the mold and could cause internal damage such as a drop.

The jack clamp is the preferred, but most expensive clamp. Its operation is simple. The foot is placed under the bottom board and the clamping foot is slid down against the cope flask. A bar is placed in the cam lever and pushed down until the clamp is snug and tight.

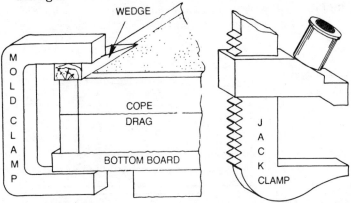

Fig. M-8. Mold clamps.

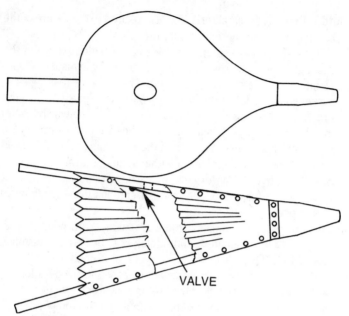

Fig. M-9. Molder's bellows.

molder's bellows: In general there are two types of bellows—one
a short snout or bench bellow and one a long snout floor bellows
(Fig. M-9). The theory is that a molder can wreck a bench mold
blowing it out with a long snout floor bellows by hitting the sand
with the snout. This theory is true due to the different stance
and angle when blowing a mold bench high or on the ground.
With care one only needs a 9 inch or 10 inch floor bellow to blow
out cope and drag molds, sprue hole and gates.

molder's blow can: The blow can is a simple mouth spray can
used to apply liquid mold coats and washes to molds and cores, or

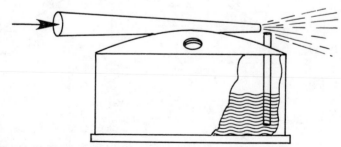

Fig. M-10. Molder's blow can.

with water to dampen a large area (Fig. M-10). It can also be operated with the air hose like the sucker.

molder's shovel: A shovel 38 inches long with a 9 inch by 12 inch flat blade. The handle is wedge shaped into a peen and a rubber piece is dovetailed into a cast handle (Fig. M-11).

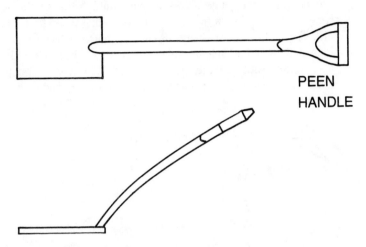

PEEN
HANDLE

Fig. M-11. Molder's shovel.

mold anchor: Mold anchors are used to prevent the shift between the cope and the drag of molds made in slip or snap flasks (Fig. M-12). After the pattern has been drawn, but before the mold is closed, one end of the mold anchor is pressed into the sand of the drag and the cope will close on the other end, securely locking the cope and drag in perfect alignment. Mold anchors can be purchased or made. They are made of sheet steel or tin plate.

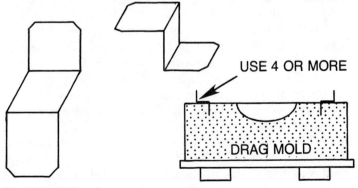

USE 4 OR MORE

DRAG MOLD

Fig. M-12. Mold anchor.

mold hardness gauge: A spring-loaded dial indicator gauge used to determine the mold hardness (Fig. M-13). The gauge is marked off from 0 to 100. The gauge measures the depth or amount of depression a convex shaped plunger makes when pressed down on a mold surface reading in mold hardness.

In operation the gauge is pressed down until the base until the base anvil is level with the sand and the reading taken from the dial. The mold hardness gauge has a brake button which freezes its action. Readings may be taken in a deep pocket, the brake applied then withdraw the gauge and read. An excellent tool to not only gauge mold hardness but locate soft spots and uneven ramming.

Fig. M-13. Mold hardness gauge.

PLUNGER

molding board: The molding board is a smooth board on which to rest the pattern and flask when starting to make a mold (Fig. M-14). The board should be as large as the outside of the flask and stiff enough to support the sand and pattern without springing when the sand is rammed. One is needed for each size of flask. Suitable cleats are nailed to the underside of the board. Their purpose is to stiffen the board and to raise it from the bench or floor to allow you to get your fingers under the board to roll over the mold.

mold shift: A shift or misalignment between the cope cheek and or drag sections of a mold causing a likewise shift in the casting.

mold weights: All snap flask work is weighted before molding. In most cases all that is needed is a standard snap weight which is

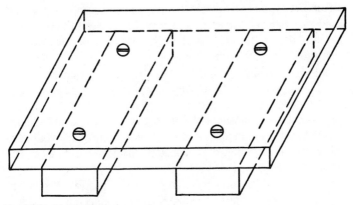

Fig. M-14. Molding board.

usually cast by the foundry to suit their own snap sizes (Fig. M-15).

The snap flask mold weights are from 1½ to 2 inches thick cast iron and weigh from 35 to 50 pounds for a 12 × 16-inch mold. The rounded cross-shaped opening through the weight is to accommodate pouring the metal into the mold. The weights are always set so that the pouring basin is free to easily see and pour. It should not be too close to the weight.

In some cases two or more stacked weights are used per mold.

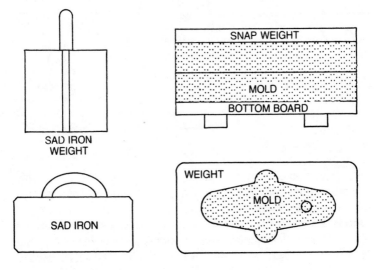

Fig. M-15. Mold weights.

Another type of mold weight which I favor for snap and small rigid flask bench work is the sad iron type of weight. They look like the irons heated on the stove for pressing clothes, except that they are quite a bit heavier.

molybdenum: A silvery-white metal, Mo, occurring chiefly in the mineral molybdenite and as a by product from copper ores. Its widest foundry use is steel to reduce grain growth. In cast iron, it inhibits the decomposition of Austenite to produce a strong fine grained iron. In both steel and iron it is commonly alloyed with other elements as nickel, chromium, manganese, copper and vanadium.

monel: A natural alloy produced directly from Canadian Bessemer Matte by the reduction of nickel ore, or produced by alloying. The average composition is 67 percent nickel, 28 percent copper with the remaining balance of 5 percent iron, manganese and silicon. Its melting point is approximately 2460°F. It can be cast, rolled or forged. 65,000 PSI tensile up to 100,000 PSI (cast) 50 percent elongation.

motor chaplets: A riveted head. Round or square-headed chaplet. Some have a round head on one end and square head on the other end. Others have a sprig or peg on one ond to locate it in a hole in a core (Fig. M-16).

 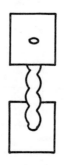 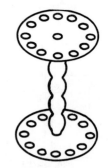

Fig. M-16. Motor chaplets.

mottled cast iron: Cast iron consisting of a mixture of variable proportions of gray cast iron and white cast iron. It has a mottled white and gray fracture.

mounted pattern: When a pattern is mounted to a board to facilitate molding, it is called a mounted pattern. In this case the mount guides on each end which match up with the flask used to make the mold. The plate is placed between the cope and drag

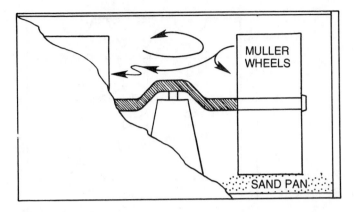

Fig. M-17. Muller machine.

flask. The drag is then rammed and rolled over. The cope is now rammed and lifted off. The plate with the attached pattern is lifted off of the drag half. The mold is then finished and closed.

muller: A sand mixing machine which conditions and mixes the sand by the mulling action of one or more large flat face wheels rotating in a tub, but not touching the bottom (Fig. M-17). Plow direct or wind row the sand in front of the mulling wheels which mull the sand with a smearing action as they roll over the sand. Some mullers are constructed where the tub rotates and the wheels are stationary. Others have the tub stationary but the wheels travel in a circle.

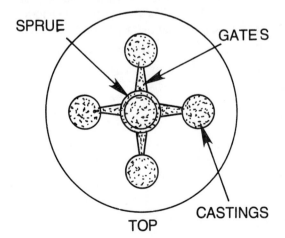

Fig. M-18. Multiple molds.

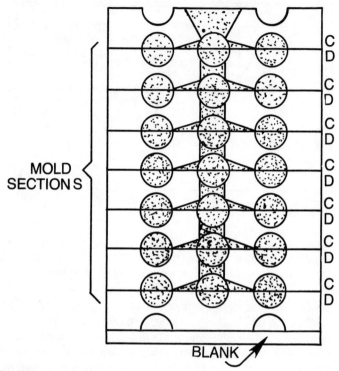

Fig. M-18. Multiple molds.

multiple molds: A composite mold made up of stacked sections. Each produces a complete gate of castings and is poured from a single sprue (Fig. M-18). The top of one section forms the drag of the section above it and the cope of the section below it.

N

natural molding sands: Contains from 8 to 20 percent natural clay. The remaining material consists of a refractory aggregate, usually silica grains.

Any natural sand containing less than 5 percent natural clay is called a *bank sand* and is used for cores or as a base for synthetic molding sand.

Commercial molding sands mined by various companies usually acquire the name of the area where they are mined. The most popular naturally bonded molding sand is called Albany. It is mined in several different grades by the Albany Sand & Supply Co., Albany, N.Y. The origin of this sand is from the pleistocene ice sheet of approximately 20,000 years ago which swept down from the north and completely overran what is now known as the Albany District. The result after eons is a seam of fine molding sand approximately 15 inches thick directly under an overburden of 8 inches of top soil.

neck down cores: A core used to neck down a riser at its point of contact with the casting to aid in its removal. In turn, it defeats the action of the riser to some extent as the neck is first to freeze (Fig. N-1). Long skinny necks on risers render the riser useless.

neutral refractories: A refractory which is neither definitely acid or definitely basic. Chrome refractories are the most nearly neutral of all commonly used refractories. Of course you have changes at high temperature due to chemical reaction, thus the use of the term *neither definitely.*

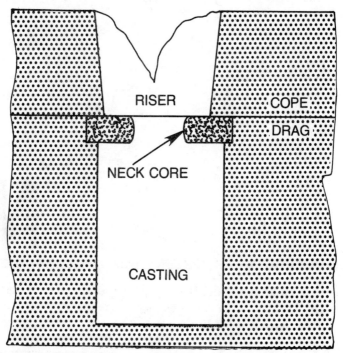

Fig. N-1. Neck down core.

nichrome: An alloy of nickel and chrome used to produce electric heating elements and parts subjected to high temperatures, kiln furniture, etc.

nickel: A silvery-white metal with a yellowish cast isolated in the year 1751. It has been used in alloy with copper since ancient times. Specific gravity of 8.84, melts at 2646°F. Widely used as an alloy in both ferrous and nonferrous metals. The uses of nickel could fill many a volume, nickel brass, nickel cast iron, nickel steel, nickel bronze, nickel chrome alloys, nickel plating, etc.

nitriding: The process of case hardening where a ferrous metal is heated in contact with a nitrogenous material or in an atmosphere of ammonia. It produces surface hardening by the absorption of nitrogen.

nodular graphite: Free graphite in cast iron in the form of nodules or spheroids (nodular iron).

nonsilica parting compound: A dry parting containing no free silica usually made of a high melting wax powder.

normalizing: Heating a ferrous alloy casting to above its transformation temperature and allowing it to cool slowly at room

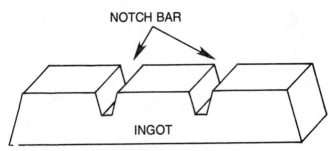

Fig. N-2. Notch bar facilitates breakage.

temperature in still air. Reduces internal stresses. Also called *stress relieving*.

normal segregation: Concentration of alloying constituents that have low melting points in the portion of the casting that solidifies last.

notch bar: Small size ingots with notches to facilitate breakage (Fig. N-2).

nowel: A colloquial term for the drag mold, not often used except by old timers.

nozzle opening: The opening in the bottom of a Bessemer ladle through which the metal is teemed when the nozzle stopper is lifted.

nugget: A small mass of metal such as silver and gold found in nature.

O

odd side: A flask or frame used simply to support a loose pattern to establish an approximate parting during the ramming of a cope or drag, after which the mold being made is rolled over and the odd side is removed and shaken out. It is replaced with a cope or drag section and the mold is completed as usual. The odd side is often referred to as a *false cope* or *false drag*.

offset matchplate: The area inside of the flask is offset which produces a matching male and female offset in the completed mold which by its locking or mating nature helps to eliminate mold shift when producing snap work (Fig. O-1).

The sure lock matchplate by Kindt Collins, Cleveland, Ohio accomplishes the same thing as an offset plate only much more accurately by producing a series of male and female matching projections and sockets around the inside perimeter of the flask.

oil no-bake core and mold process: A synthetic oil binder mixed with basic sands and chemically activated to produce dri-sand cores and molds at room temperature. It is an acid dehydration or condensation reaction. The oxidation reaction is accomplished chemically rather than by heat and oxygen.

oil oxygen process: Used to produce dri-sand cores and molds by the principle of auto oxidation. The system utilizes both oxidation and polymerization mechanics on a combination of oils containing chemical additives so that the final product is activated with oxygen-bearing materials. It will set (harden) in a predetermined time.

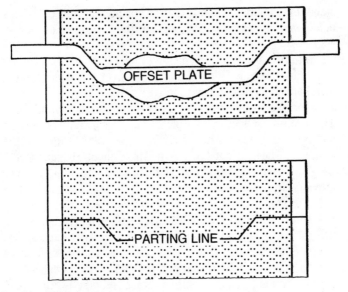

Fig. O-1. Offset matchplate.

old casting for pattern: In some cases a broken or old casting can be used as a pattern to produce a new or additional casting. In the case of a simple casting which leaves its own core and has ample draft like a cast iron frying pan, this can be a cinch. Large gear wheels have been cast from a broken gear as a pattern by waxing draft on the teeth, plugging the bore through the hub with a core print and rapping the gear heavily in the mold to take care of the shrinkage. Old machinery replacement parts are often produced this way.

olivine foundry sand: A translucent mineral—Mg.Fe. 2.SiO₄. A solid solution of forsterite and fayalite. Also called *chrysolite*. The choice green stones are used as gems, peridot. Olivine sand is highly refractory. The melting point of forsterite is 3470°F. It is expensive, but an excellent molding sand (when bonded with bentonite) and a great core sand for the tough jobs. *Olivine flour.* Olivine sand ground to a flour makes an excellent investment mix—two parts plaster of paris, three parts olivine flour. Also excellent core and mold washes bonded with molasses water or goulac water.

omission of a core: Results and cause obvious.

open grain structure: A defect where the grain is too coarse when machined or fractured. The causes are many, including carbon equivalent too high for the section, lack of carbide

331

stabilizers, pouring too hot (in iron) or pouring too cold. This defect can be completely throughout the casting or in random spots.

open hearth: A furnace for melting metal in which the bath is heated by the convection of hot gasses over the surface of the metal and by radiation from the roof.

open sand casting: Casting in an open sand mold with no cope. This process is used to cast plates for rigging and where large weights are needed. The big trick in open sand work is in the ramming. The sand must be rammed hard enough to support the weight of the metal; otherwise, without the hydrostatic pressure provided with a cope and sprue height, the metal will kick badly and you wind up with a dangerous mess.

It is very tricky to say the least. It requires much greater skill than molding with a cope. Pouring an open sand mold is also tricky. The trick is to pour from one side, starting a wave of metal across the mold, slacking off and then pouring again in such a way that you catch the returning wave (rebound) with second wave. You repeat this action until the mold is poured.

optical pyrometer: A temperature measuring device through which the observer sights the heated object and compares its incandescence with that of an electrically heated filament whose

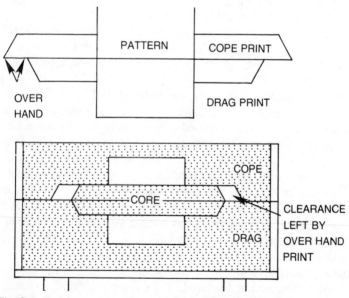

Fig. O-2. The over hand provides clearance for closing the mold.

brightness can be regulated. The operator regulates the brightness of the filament until it matches by disappearing from the screen. The brightness of the object then reads the temperature in F or C on the side of the instrument.

optimum moisture: That moisture content which results in the maximum development of any property of a sand mixture.

organic material in melting: The reference here is to the gassing and inferior properties of castings (mainly nonferrous) where the scrap, chips, borings, etc., are not completely cleaned of cutting compounds and oils.

over hand: The extension of the end surface of the cope half of a core print beyond that of the drag in order to provide clearance for closing the mold (Fig. O-2). It prevents the cope from shaving if not closed completely level.

over iron: The amount of iron left in a ladle or excess melted for the floor. This over iron must be kept to a minimum but at the same time it must be short with the heat. In a ladle it can be more costly. It is similar to pouring a 500-pound casting with 450 pounds of metal in the ladle.

oxidation: Any reaction whereby an element reacts with oxygen. Oxidizing the heat from the products of combustion, rust, etc.

oxygen lance: Simply a length of ⅛ to ¼-inch steel pipe connected by a flexible hose to a regulator and oxygen bottle. It is used to burn through steel or burn through a frozen tape hole on a cupola.

P

pad: A shallow projection on a casting distinguished from a boss or lug by shape and size only (Fig. P-1).

paddle mixer: A sand mixer which mixes the sand with a series of offset metal paddles in a churning action, as a plaster mixer.

parted pattern: A pattern made in two or more parts (Fig. P-2).

parting: The zone of separation between the cope and drag portions of a mold, flask or pattern. Also called *parting plane*. The word parting also refers to any material used to effect a parting, parting powder, PVC, parting compounds, etc.

parting compound: A material dusted or sprayed on patterns to prevent adherence of sand and to prevent the sand to sand joint of a mold from adhering together.

parting line: A line on a pattern or casting corresponding to the separation between the cope and drag portions of a sand mold (Fig. P-3).

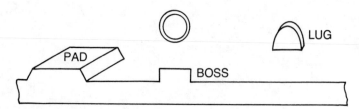

Fig. P-1. Pad.

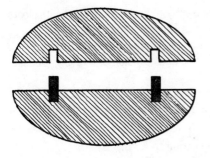

Fig. P-2. Parted pattern.

pasted cores: Cores which are made in halves and after they are dried, glued together to make the complete core. If the core is symmetrical, a half box is all that is needed.

After the two halves of the core are dried, a vent is scratched along the centerline of one half and the sections are glued together with core paste. The seam is then mudded with a material known as *core daubing*. Both core paste and daubing can be purchased or made. For homemade core paste, use enough wheat flour dissolved in cold water to produce a creamy consistency.

For a good core daubing, mix fine graphite with molasses water—one part molasses to 10 parts water. Enough graphite is added to the molasses water to make a stiff mud.

Another good daubing mix is graphite and linseed oil mixed to a stiff mud.

After a core has been pasted and daubed, it is a good idea to return the core to the core oven for a short period to dry the paste and daubing.

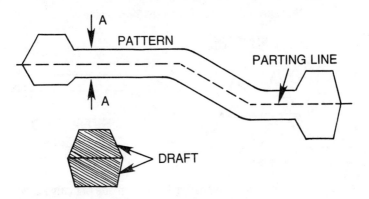

Fig. P-3. Parting line.

pattern board: A true surface smooth board reinforced with cleats upon which a loose pattern is laid for ramming the drag.

pattern coating: Coating material applied to wood patterns to protect them against moisture and abrasion of molding, shellac, etc.

pattern colors (wood patterns):

Patterns are colored with colors to indicate the following:

- Black—Surface to be left as cast (unfinished).
- Red—Surfaces of the casting to be machined.
- Yellow—Core prints and seats of and for loose core prints.
- Red stripes on a yellow background—Seats of and for loose pattern parts.
- Black stripes diagonally on a yellow background—Stop-offs.

By looking at a properly colored pattern the molder can determine if he has all the parts, cores and exactly how to proceed with the mold making.

pattern letters: White metal, brass or aluminum letters and figures for attaching to wood or metal patterns for identification, name, directions, etc (Fig. 9-4). The letters can be purchased in a wide variety of letter styles and sizes from as small as 3/32 inch to 4 inches in size. They can be had plain back, with sprigs, drilled or with reverse letters for making branding irons. They are used widely to make grave markers, plaques, signs, etc.

Plain back letters are usually glued to the pattern with glue or shellacked. Sprigged letters are attached to wood patterns by tapping them down in place with a soft hammer. The sprigs act as brads. Drilled letters are pinned or screwed on to the pattern. The letters are all drafted correctly for molding.

DRAFTED

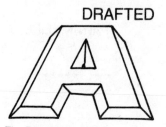
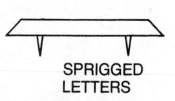

SPRIGGED
LETTERS

Fig. P-4. Pattern letters.

pattern lumber: Although patterns can be made of many kinds of woods, most wood patterns are made of mahogany or sugar pine. Mahogany is usually used for fine delicate wood patterns but is not confined to small patterns. Mahogany, Philippine, African, Mexican, Honduras, Peruvian and Cuban are available for use to the patternmaking trade; however, Mexican, Honduras and the Peruvian grades are by far the most used for pattern work. It is a hard wood easily carved and worked and will withstand the hard use and abrasion encountered in the foundry exceptionally well. It weighs approximately 3 pounds per board foot and can be purchased in random lengths and widths from 4/4 (1 inch thick) to 16-4 (4 inch thick) in various grades. The better select grades are used for pattern work.

Sugar pine, a native wood of California and southern Oregon, weighs 2.3 pounds per board foot. It is a softer and lighter wood than mahogany and considerably cheaper. However, it is widely used and holds up extremely well under use in the foundry. It's also available in a wide range of widths, lengths and thicknesses. The select grades are the most used for patterns.

For years Jelutong wood was used throughout the rest of the world for pattern work and it is now becoming popular in the western hemisphere. It is a white or straw colored wood with no differential between heart wood and sapwood, a moderately fine even texture, straight grain and low luster with a weight of pine.

penetration: A casting rough surface defect which appears as if the metal has filled the voids between the sand grains without displacing them. It is caused by a too coarse molding sand for the metal and casting weight.

perforated chaplets: Tin coated sheet steel chaplets made from perforated stock (Fig. P-5).

perforated tin sheets: Perforated tin plated sheets of steel used to form various types of skimmers and gates, sold in a wide variety of thicknesses and perforations.

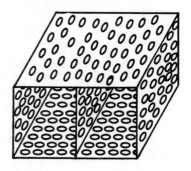

Fig. P-5. Perforated chaplet.

337

permeability: The ability of the mold material to allow the steam to pass through the walls. Permeability can be measured with a meter which measures the volume of air that will pass through a test specimen per minute under a standard pressure. Some instruments are designed to measure a pressure differential which is indicated on a water tube gauge expressed in permeability units.

permanent molds: A mold made of metal or graphite. It is used repeatedly for the production of many castings of the same form, not an ingot mold. It is usually gravity poured. Ninety-nine percent of all automotive pistons are cast in permanent molds.

Petro Bond: Petro Bond is a patented oil bonding system of NL Baroid Industries. Petro Bonded sand consists of a fine silica and with a maximum of ½ percent of clay (preferably clay-free) mulled with petro bond oil, Petro Bond powder and P-1 catalyst. Typical green sand properties using a GFN consists 140 sand, compression strength 12 psi, permeability 15, flowability 87, rammed hardness (mold hardness gauge) 84 and a green deformation in inches .010 to .014. It is basically used as nonferrous green sand and with excellent results up to and including a 4500-pound bronze casting. Two pounds of iron oxide per 100 pounds sand is beneficial to the correct balance between fine and coarser sand. If Petro Bonded sand is used as a facing sand, it will mix in readily with your flour or system sand with no problem.

It has been my personal experience with petro-bonded sand that if just sufficient graphite is added to give it a gray color (2 percent to 3 percent), the results are superior to those with just the iron oxide. It is not nearly as sticky, has excellent lifts, smoother casting surface, the flowability is increased 100 percent and the mold hardness is improved. Although you need a muller to prepare the initial batch I find it can be kept in shape for months on end by simply riddling it with a *gyratory riddle*. Most foundry supply houses will supply you with petro bonded sand ready to use (mulled by them).

Although it is no panacea for small hobby shops, for schools with little or no sand conditioning equipment, it is great. Due to the absence of moisture a finer sand can be used and the problems caused by moisture are eliminated. Liquid parting cannot be used. Use a good grade of commercial dry parting or use petro bond (the dry binder) as a parting.

Some large commercial foundries have converted to complete Petro Bonded systems. A typical Petro Bond mix consists of 100 pounds dried silica sand, 5 pounds Petro Bond binder, 2 pounds Petro Bond oil, 1 ounce catalyst P-1 or 100 pounds dried silica sand, 5 pounds Petro Bond binder, 2 pints metro 20 (oil) from Mobile Oil Co. and 1 ounce methanol alcohol.

Petro Bond[(R)] is a registered TM of NL Industries Inc.

The mixing procedure for Petro Bond sand is divided into three phases:

- First: Weigh out 100 pounds of sand and mix into it 5 pound of Petro Bond. Dump the load into the muller and mix dry for about one minute.
- Second: Add to the mix in the muller 2 pound or 2 pints of Petro Bond oil and mull for about 10 minutes.
- Third: Add 1 ounce catalyst P-1 and mix for 3 to 5 minutes longer. With many sand the green strength will be at least 8.5 psi. The green strength required will depend on the type of metal to be cast and the size of the casting to be made. After mixing, the sand can be kept indefinitely.

Mixing. The sand mixture should be mixed in a sand muller. The time of mixing varies with the type of muller used and can be determined by varying the mixing time until the desired strength is obtained.

In a slow muller the dry ingredients should be mixed for a minimum of one minute. Subsequently, the oil can be added slowly while the muller is in motion and mulled for a period of 10 minutes. The catalyst P-1 is then added and mixed for 3 to 5 minutes. The mix can then be stored indefinitely and will be usable at any time without further treatment.

If this mix results in a stronger sand than desired, it may be cut with clean sand.

As the sand is used it may become contaminated with coarse particles of core sand. If this contamination proceeds to the point where the over-all percentage of fines in the sand is noticeably reduced, it is advisable to add iron oxide during one of the remulling cycles. Such additions help restore the correct balance between fine and coarser sand. The usual procedure is to add the iron oxide at the rate of from one to two pounds per 100 pounds of sand. This addition will also result in a tougher but drier sand. It may then be necessary to add additional oil to restore proper moldability.

When it is used for the entire mold, after the casting has solidified and is *shaken out*, it is only necessary to *aerate* or *riddle* the sand before re-using. This sand can be used without remulling until the green strength has been reduced sufficiently to cause scabbing or washing.

Petro Bond sand may be used as a facing for green sand molds. When Petro Bond sand is used as a facing, the burned sand readily mixes with the green backing sand.

For both types of use, best finishes are obtained when the mold hardness is 80 or higher.

Equipment. Patterns made of wood, plaster, aluminum, brass, steel, etc., can all be used. Bear in mind that the accuracy and finish of the casting produced can be no better than the pattern. Waxed patterns should not be used until they have been coated with a hard resin or paint.

Troubleshooting. The following difficulties usually can be eliminated by using the procedures described. If the trouble continues, carefully check all materials, equipment and procedures being used.

- Low Green Strength:

The minimum green strength of 8½ pound should be obtained if the mix is adequately mulled.

Improper milling—The formula may be correct and the ingredients good, but the actual mulling achieved may be inadequate. The muller must be clean and dry and have a rough enough surface to provide intensive mixing. The wheels of the muller should be lowered to the pan to give the proper mulling action. Plows that are too worn will not provide proper mixing. If portable mulling equipment is being used, the time of mulling should be greatly increased (above the time required for mulling by heavier stationary equipment) in order to get a mix of proper green strength.

Petro Bond content too low—Check the amounts of materials being used in the mix. Make sure you have used the amounts and the ratio of Petro Bond to oil that is prescribed in the foregoing mixing procedures. Be sure the Petro Bond is uniformly dry mixed and the oil is added slowly to avoid leakage.

Too much moisture in sand—Check the sand to make certain that the moisture content is less than ¼ percent.

Oil is not of the proper type—Recheck the oil being used. Make certain that the oil is of the proper specifications. If the finished mix has a glossy appearance, it indicates that the oil is probably not of the proper specifications.

● Poor Finish of Castings:

Incorrect mold hardness—Check your mold hardness. Make sure that its hardness is 80 or better. If you are using a very high green strength in your mix, make certain that the mold is rammed properly to give you a minimum mold hardness of 80.

Wrong parting agent—Check on the amount and type of parting agent being used. *Do not use a liquid parting agent with Petro Bond.* A dry parting is necessary if loose patterns are used, especially if they are wooden patterns that have been shallacked. Petro Bond can be used as a dry parting.

● Turbulent Metal Flow or Lack of Effective Choke Feeding:

To assure smooth metal flow and adequate filling of the cavity, it is necessary to choke the flow of metal to eliminate turbulence. Most metals can be poured at lower temperatures due to the absence of the chilling effect of water and the lower heat conductivity of the oil.

● Cutting and Washing in Molds:

Improper ramming of the mold—Make certain that the mold is rammed hard. Check the ingates and sprues to make sure they are properly cut. If the gates and risers are cut after the mold has been made, it is advisable to lightly swab the surface of the cut with the same oil used in the Petro Bond mix. This helps to bind the grains of sand and minimize any washing into the casting. Ingates and sprues should be smaller in the Petro Bond molds than in conventional sand molds.

Improper pouring of molds—Petro Bond sand molds can be poured at considerably lower temperatures. If the surface of the Petro Bond mold is very smooth, the flow of metal should be reduced or restricted. This can be accomplished by using strainer cores or by changing the gating system so that the metal enters the mold without turbulence.

● Non-Uniform Reproduction of Pattern:

Improper mold hardness—If it is found that the cope and drag surfaces of the castings are smooth but the side walls are not, it is an indication that the mold has not been rammed hard enough. The mold hardness should be uniform on all the mold surfaces. Cope, drag and matchplate equipment give the best reproduction.

● Out of Dimension Castings:

Insufficient cooling time—When castings are removed from the mold too soon there is a tendency of the casting to warp because of non-uniform solidification. It is essential to leave the casting in the sand long enough to have it properly solidified.

●Gas Problems—Blows, Cold Shuts, Etc.:

Too much oil in the mix—When excessive amounts of oil are used in Petro Bond mixes, the surface of the metal may show slight imperfections. To eliminate this condition, mull 2 percent iron oxide into the mix to absorb the excess oil. Another method is to rebalance the formula by the addition of clean silica sand and Petro Bond to bring the mix to the proper green strength.

Permeability should be checked—Molds made with Petro Bond are poured at considerably lower permeability than molds made with water-bonded sand. There is, however, a minimum level of permeability. This is particularly true when pouring high temperature alloys. The optimum must be established in order to give proper castings.

●Poor Castings Finish—Result of Use of Same Sand Over a Period of Several Months:

Excessive contamination by core sand—Check your system and see if excessive amounts of core sand are being introduced.

phosphorous: A nonmetallic element, P, widely diffused in nature and found in many rock materials and ores. In the foundry it is used as a deoxidizer combined with copper or other alloys to stabilize it so that it can be handled.

phosphoric acid: H_3PO_4 used in pickling and rust proofing. Widely used as the catalyist in the Furan no-bake sand system.

pickle: To clean metal by chemicals or electro-chemicals. To remove surface oxides, etc. An acid or caustic solution.

pig bed: Small scooped-out pig shaped excavations made in an open sand mold or bed to pour off the excess metal from a heat.

pig iron: Cast iron produced by the reduction of iron ore in a blast furnace.

pinch dog: A steel C-shaped device with sharp points which are tapered on their inner face so as to give a clamping effect when driven into wood (Fig. P-6). The further in, the tighter the clamp (pinching) effect. Used in pattern work to hold one member to another while gluing, nailing or drilling. The prime use is in gluing up stock. Some patternmakers use them to hold the two halves of stock together when turning a split pattern. This is a very dangerous practice. If they become loose while turning, the centrifugal force will throw them with great impact. You could be seriously injured or killed.

Pinch dogs come in sizes from ½ inch to 3 inches between points. Purchased from any pattern supply or foundry supply company they can be used for a variety of tasks.

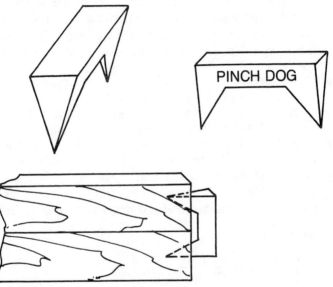

Fig. P-6. Pinch dog.

pin holes: Surface pitted with pin holes which may also be an indicator of subsurface blow holes.

pine tree structure: The appearance of dentrites crystal growth in a solidifying metal.

pitch and gilsonite: Both of these materials are used to produce dry and hot strength along with good hot deformation and as expansion controls. Also used as a binder for dry sand work and black sand cores. It is a baking type of binder. Most prepared black core binders sold today under various trade names are usually a mixture of pitch, destrin, liglin sulphite and resin. In order to use these correctly you should know the percentage of pitch. Light work pitch is used by percent by weight. When using pitch in dry sand work it must be worked well on the wet side. It is used in skin dry and dry sand work for large aluminum bronze work and dried overnight. Sometimes it is used in the combination of 50:50 with sea coal and also with southern bentonite. When used as a replacement to sea coal, it can be used as low as ½ percent to as high as 4 percent by weight. The mold hardness, toughness and green strength increases as does the hot and dry strength. The permeability decreases somewhat and little or no effect on the flowability is encountered. It should be noted that other materials are also used as ingredients in facings such as liglin sulphite, rosin, graphite, coke and fuel oil. Gilsonite is a natural pitch much more potent than coal tar pitch and usually is used sparingly in the order of ½ to 1 percent in green sand facing mixes. It is also used in the manufacture of some mold and core washes.

porosity: Unsoundness of a casting due to the presence of blow holes, gas holes or shrinkage cavities.

poured short: Casting incomplete due to not filling the mold. Going back and touching up the mold will not do it.

pouring basin: A basin on or in the cope to hold the metal prior to its entrance into the sprue.

pouring devices: Any device that is used to handle ladles or crucibles for pouring molten metal. They vary from simple shanks to complex apparatus and automatic robodic devices.

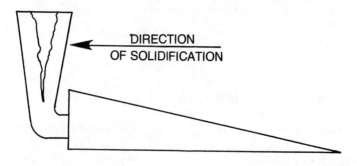

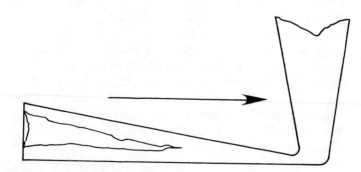

Fig. P-7. Progressive solidification.

pressure: Too low a pressure (hydrostatic) by having too shallow a cope flask. Thus a short sprue will cause scars, seams and plates as a casting defect.

pressure tight: A casting that does not leak under pressure. It is free from porosity and unsoundness.

primary crystals: The first dentritic crystals formed in an alloy during cooling below the liquidus temperature.

print back: Where the pattern is removed from the mold and the sand cavity dusted with cement or plumbago. The pattern is returned and rapped to set the surface and redrawn. This system is used to produce an extremely smooth casting surface. It is widely used when casting plaques and grave markers.

progressive solidification: The solidification of a casting from the thin sections the farthest away from the sprue toward the sprue or riser (Fig. P-7). It is accomplished by correct design, chills, correct gating and risering.

prop: The iron post that holds the cupola bottom doors closed.

protein binders: Core and mold binders which are organic compounds containing nitrogen.

pull cracks: Hot tear.

pull down: A sand buckle in the cope due to excessive expansion of the cope sand. Insufficient combustibles in the sand (wood flour, etc.).

push-up: An indentation in a casting usually on the drag side due to the displacement of the sand in the mold. Often caused by improperly seated bottom boards, clamping practice or too shallow of a drag for the job.

quenching: Heating a casting to a specified temperature and quenching it in water, oil, etc., to create a heat treatment.

R

radiant heat: The heat radiated from a furnace arch and walls. Also the heat radiated from the molten metal poured into a mold. The cause is often due to metal penetration due to the improper location of gates and risers which promote localized overheating of the sand. Pouring a mold with an insufficient number of gates will subject the mold surfaces to radiant heat for too long of a period and will cause the mold surface to break down. A very common problem with investment castings which are undergated and prevent the mold from filling fast enough.

radiator chaplets: Differ from most other chaplets in that they are set before the mold is rammed (Fig. R-1). The practice is to drill a hole in the pattern corresponding to the diameter of the wire used in the chaplet. The chaplet is dropped into the hole, sand rammed into the mold and the pattern withdrawn. The stem of the radiator chaplet is left protruding in the mold cavity a distance equal to the thickness of metal and ready to support or hold down the core in its proper position.

When the core is set, it rests against the flat end of the wire stem. In many cases this does not provide a sufficient bearing surface when the casting is poured. Additional bearing surfaces can be obtained by putting a round or square core plate on the core, against which the radiator chaplet will press.

The most commonly used radiator chaplet has a shoulder and break-off nicks. This drops into hole in the pattern as far as the shoulder.

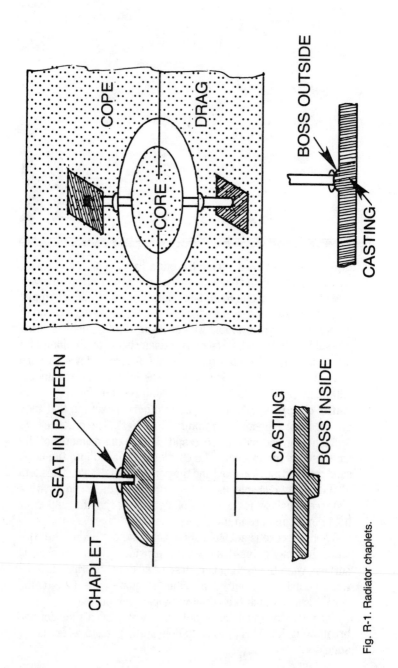

Fig. R-1. Radiator chaplets.

ramaway: A defect caused by molding with a sand that has poor flowability and ramming at the wrong angle (Fig. R-2). A section of the mold being forced away from the pattern by ramming sand after it has conformed to the pattern contour.

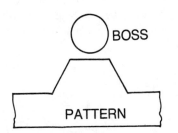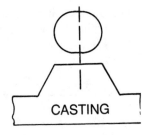

Fig. R-2. Ramaway.

ram off: A defect resulting from a section of the mold being forced away from the pattern by ramming sand after it has conformed to the pattern contour. This is caused by careless ramming where the mold is rammed vertically and then on an angle, causing the vertically rammed sand to slide sideways leaving a gap between the pattern and the sand. It results in a deformed casting. Another cause is using a sand with poor or low flowability.

ram-up core: A ram-up core is a core that is set against the pattern or in a locator slot in the pattern (Fig. R-3). The mold is rammed and when the pattern is drawn the core remains in the mold.

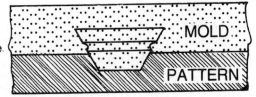

Fig. R-3. Ram-up core.

rapping bar and rapper: The rapping bar consists of a piece of brass or steel (cold roll) rod which is machined or ground to a tapered point (Fig. R-4). The rapper is made of steel or brass and is shaped exactly like the frame of a sling shot.

The purpose of these tools is to rap or shake the pattern loose from the sand mold in order to draw it easily from the sand. What you are doing is shaking the pattern in all directions which will drive the sand slightly away from the pattern. The resulting mold cavity will actually be a fuzz larger than the pattern. The pattern should only be rapped enough to free it from the sand. You will be able to see when it is loose all around by the movement of the pattern. Over rapping will distort the mold cavity which may or may not matter depending on how close you wish the casting to hold a tolerance.

The operation is quite simple, the bar point of the rapping bar is pressed down into a dimple in the parting face of the pattern with the left hand. The rapper is used to strike the bar with the inner faces of the yoke.

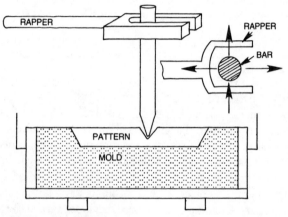

Fig. R-4. Rapping bar and rapper.

rat: A lump on the surface of a casting caused by a portion of the mold face sticking to the pattern.

rat tail: A minor casting defect. A small buckle occurring as a small irregular line or lines on the casting surface.

receiving ladle: Usually a ladle (often stationary in front of the cupola) into which the metal is tapped and where any innoculation needed is done. The pouring ladles are fed from the receiving ladle. In this arrangement the receiving ladle becomes in fact a tilting forehearth.

red brass: Red brass is defined as a brass alloy containing 2 to 8 percent zinc with the tin content less than the zinc. The lead content is less than .5 percent. This alloy is not often used in the production of castings. The leaded red brasses and semi-red are called red brasses. A leaded red brass would be a brass with 2 to 8 percent zinc with a tin content less than 6 percent (usually less than the zinc content) and lead over .5 percent. 85 percent copper, 5 percent tin, 5 percent zinc and 5 percent lead is called red brass but is actually a leaded red brass.

red shortness: The brittleness of some ferrous and nonferrous metals when it is at a red heat. Referred to as *hot shortness*.

reducing atmosphere: An atmosphere which is short or lean of oxygen.

refractoriness: The ability of sand to withstand high temperatures without fusing or breaking down.

regenerative chamber: A system where the air for combustion is pre-heated by regenerative chambers which are heated brick maze through which the blast air passes and is heated. They are constructed in pairs while one is giving up its heat to the blast air and the other is being heated (usually by the exhaust gasses from the melting unit). When the chamber through which the blast air is being heated falls below a certain temperature, a series of valves swap the direction of the air and exhaust over to the other chamber to regenerate the cooled one.

reservoir: A basin on top of a mold designed to hold sufficient metal to pour the mold. It is opened to the sprue by a plug, similar to a bottom pour ladle (Fig. R-5).

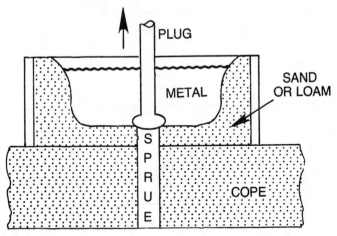

Fig. R-5. Reservoir.

return scrap: Home generated scrap consisting of gates, risers, chips and borings from the castings produced. In some cases as with steel, the return scrap from a given casting can be greater or equal to the weight of the casting itself.

reverse chill: Also called *inverse chill* (Fig. R-6). The condition in a casting section where the interior is mottled or white iron with a gray iron exterior. It is usually found in thin castings. The usual

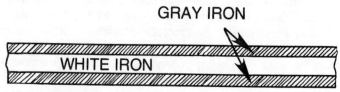

Fig. R-6. Reverse chill.

cause is the presence of nonferrous metals in the iron such as antimony, lead, tellurium, etc., or the casting was poured with a too high carbon equivalent for the section thickness being poured.

ribs: Stiffening members on a casting designed to increase its strength and or utility.

riddle (hand): A round riddle or seive used by a hand molder to riddle sand or facing on to a pattern prior to ramming the mold. Riddles are purchased with various meshes of wire. The most common hand riddle used is a #4—four openings per inch.

risers: A reservoir or reservoirs of excess molten metal designed to supply metal in compensation for the shrinkage that cannot be properly fed from the gate (Fig. R-7).

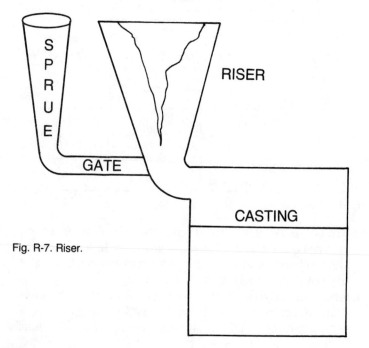

Fig. R-7. Riser.

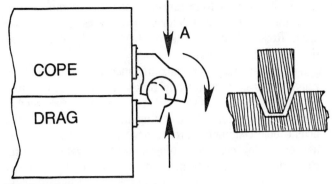

Fig. R-8. Roll-off hinge.

roll-off hinges: These hinges allow the flask to be opened like a book (Fig. R-8). The pattern is removed and the mold closed—all without lifting the cope.

rollover machine: A molding machine designed in such a manner that the flask is rolled over before the pattern is drawn from the mold, as opposed to lifting the mold from the pattern (pin-lift machine).

roman joint: A joint commonly used in the assembly of the component pieces of a large statuary bronze casting (Fig. R-9). The joint is made in such a fashion that when properly done it cannot be found or located. Knife edge ridges at the mating junction are hammered down flat (cold forged) with a peen hammer and then finished with a file and scraper.

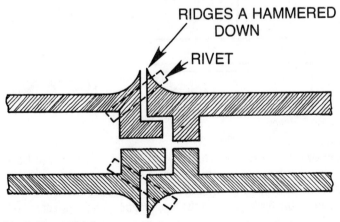

Fig. R-9. Roman joint.

rosin oil: An oil made from gum or wood rosins by the destructive distillation and then the fractionating of the distillate into an oil or spirit. Rosin oils are used as core binders alone and in combination with other materials.

rough surface: This defect can run from mild penetration to spotty rough spots or a completely rough casting. Many factors or combinations are at fault—sand too coarse for the weight and pouring temperature of the casting, improperly applied or insufficient mold coating or core coating, faulty finishing, excessive use of parting compound (dust), hand cut gates not firm or cleaned out, dirty pattern, sand not riddled when necessary, excessive or too coarse sea coal in the sand, permeability too high for class of work or the core or mold wash faulty (poor composition).

rubber cement: Latex rubber treated with a preservative and used as a core binder. The sand is mixed with latex and water and the core is formed and dried at room temperature. The latex cores have a quick collapsibility and shake out easily. An excellent binder for cores used in thin walled aluminum castings.

runner box: A device for distributing molten metal metal around a mold by dividing it into several streams (Fig. R-10).

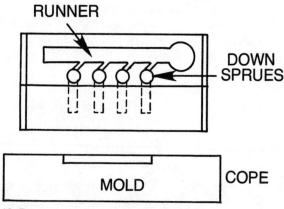

Fig. R-10. Runner box.

runner riser: A runner in the cope horizontal or vertical which acts also as a riser to feed the castings (Fig. R-11).

run out: Caused by the metal in the mold running out between the joint of the flask. It drains the liquid metal partially or completely from any portion of the casting above the parting line. A run out

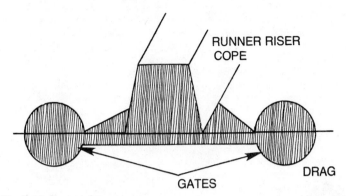

RUNNER RISER
COPE

GATES

DRAG

Fig. R-11. Runner riser.

can come through or between the drag and the bottom board from a cracked drag mold. Also, it can come from between a loose or improperly fitting core and the core print. It is caused by insufficient room between the flask and the cavity, insufficient weight on the cope (cope raises during pouring) or improperly clamped molds. Excessive hydrostatic pressure (sprue too tall for the job), no dough roll used between cope and drag (large jobs) or a combination of all of the above.

An attempt to save a run out by placing your foot on top of the cope and applying pressure above the point where the liquid metal is running out is foolhardy and dangerous. Also trying to stop off the run out flow with sand or clay is foolish.

sag: A decrease in metal section due to a core or the cope sagging. The cause is insufficient cope bars, too small a flask for the job, insufficient cope depth. This defect will also cause misruns.

sand slinger: A mechanical device which throws molding sand with sufficient force and impact so that the sand is literally rammed into the mold (Fig. S-1).

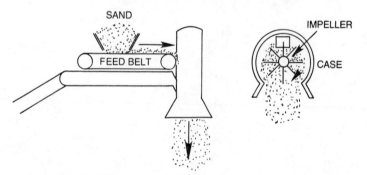

Fig. S-1. Sand slinger.

The principle is rather simple. The conditioned molding sand is fed via a belt into the hub of a high speed impeller consisting of flat blades. The blades throw or sling the sand downward through the discharge mouth. Both the amount of sand fed to the impeller and the impeller speed is controlled to give the desired impact and flow. The two basic types are stationary and a travelling slinger. The primary use is in ramming up large molds, copes and drags. They are widely used in foundries engaged in *street castings*, sewer rings, sewer covers, catch basins, etc.

The common name in the foundry is *elephant snout*. On large slingers the operator rides the business end of the slinger.

saxophone gate: A gate designed to provide the same conditions as a step gate only delivering metal to the casting at different levels as the mold is filled (Fig. S-2).

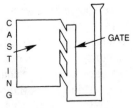

Fig. S-2. Saxophone gate.

scab: Rough thin scabs of metal attached to the casting by a thin vein separated from the casting by a thin layer of sand. Usually found on flat surface. They are caused by hard ramming, low permeability and insufficient hot strength. Sand does not have enough cushion material, wood flour, etc., to allow it to expand when heated. Unable to expand, it will buckle causing scabs along with, but not always rat tails. Grooves under the scabs are called *pull downs*. It is the pull downs that bring about the scab.

scarfing: Removing pads, gates and risers from steel castings with a hydrogen or oxyacetylene torch prior to grinding.

scrap metal: Metal purchased from a scrap metal dealer or junk yard. Also called *foreign scrap*. Scrap that is generated in the shop from the operations is called *return scrap* or *domestic scrap*. Great care should be exercised in buying foreign scrap as to its exact pedigree. It should be properly cleaned and classified. With nonferrous materials this can be quite a trick and requires a great degree of experience.

screen core box vents: Round screen vents used in a core box which is blown (Fig. S-3). The vents exhaust the air in the box from the blow. When slotted instead of screened, they are called *slot vents*.

sea coal: A finely ground soft coal used as a dry shake bag facing. Mostly used as an additive to both facing and system sand in amounts up to 5 percent. The most common practice is to keep it in the neighborhood of 3 to 4 percent. Excessive sea coal in any

SCREEN
VENT

SLOTTED
VENT

Fig. S-3. Screen core box vents.

molding sand, whether natural or synthetic, will cause the sand to become brittle, low in resilience and difficult to mold with. Care should be exercised in adding sea coal to any sand; however, properly used it is very beneficial and will produce excellent castings with good peel and a smooth surface. In the proper amounts it will increase the green strength, dry strength, temper, moisture required, mold hardness and deformation while it decreases the flowability and permeability.

The most common use in molding sand is in cast iron sands. It can be purchased in basically five grades:

● A—12 mesh for large heavy machine castings.
● B—18 mesh for medium weight castings.
● C—24 mesh for light and medium weight castings.
● D—40 mesh for radiator castings and lightweight cast ings.
● E—100 mesh for ornamental iron, piano plate, stove plate and hollowware.

The average analysis of a good grade of sea coal = volatile matter, 38.49 percent; fixed carbon, 53.84 percent; ash, 4.81 percent; and moisture, 2.70 percent. Contrary to popular belief that sea coal has no use in brass and bronze casting sands, the author has done extensive work with sea coal in every weight range from 1 pound and less to in excess of 500 pounds in brass and bronze with most gratifying results. One half to 2 percent of silk-bolted or air-float sea coal in any brass sand works very well.

Petrobond sands are improved with the addition of sea coal or graphite. The flowability and moldability is increased by 100 percent.

semi-centrifugal: When a casting is symmetrical about its own axis such as a wheel or gear blank casting and poured through its hub or center, the process is called semi-centrifugal (Fig. S-4).

semi-killed, steel: Incompletely deoxidized steel. Permits the sufficient evolution of carbon monoxide to offset solidification shrinkage.

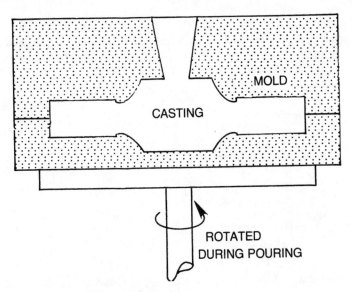

Fig. S-4. Semi-centrifugal casting.

semi-steel: A misnomer when referring to gray iron when the cupola charge consists of a high percent of steel scrap.

set gate pattern: If a pattern is made for a gate but not attached to a pattern and only placed against it while making the mold, this pattern is called a set gate pattern.

set sprue: A wooden *sprue stick* set in a sand mold during the ramming of the mold. Usually a simple, round and slightly tapered wooden stick (Fig. S-5). Some are bell shaped at the top to form a pouring basin.

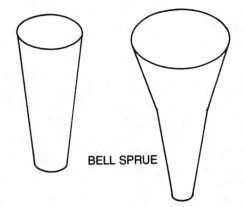

BELL SPRUE

Fig. S-5. Set sprue.

set-up core: A core print made of a dry sand core to provide a stronger bearing surface for a heavy core should a green sand print prove inadequate (Fig. S-6). A common practice in large green sand work.

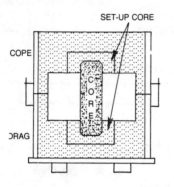

Fig. S-6. Set-up core.

shake bag: A pourous cloth bag used to shake dry parting or blacking materials on a pattern or mold surface.

shellac pot: A glass or metal pot with a domed lid whereby the brush can be left in the shellac. The cover forms a seal when closed (Fig. S-7).

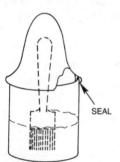

Fig. S-7. Shellac pot.

shell process: A process where a resin-coated sharp sand is dumped on a hot metal pattern and allowed a long enough time to produce a heat cured shell of the desired thickness to form shell mold (Fig. S-8). The remaining uncured sand is dumped back into the dump box and the cured shell is stripped from the pattern. The cured shells (two halves cope and drag) are assembled by bolting, gluing or clamping. They are backed with shot or sand and poured.

Cores are also made by the sand shell process with hot metal core boxes which make excellent lightweight hollow cores. They

360

are used not only in shell molds, but in sand and permanent molds as well.

When the sand shell system was first introduced into the foundry industry, it was taken by some as the panacea. Before it found its true place in the industry, it took a heavy toll not only in botched-up castings but in money.

It is a good system when properly used and its limits are understood. The same problem exists with any system. Investment casting particularly by art bronze casters is another with definite limitations. I have seen some costly messes made by investment casters who will try to cast a life-size casting all in one piece in investment when the casting could have been done in French sand with not only a finer finish but at half the time and cost. Use the right medium for the job at hand.

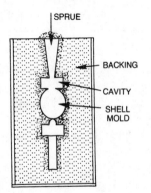

Fig. S-8. Shell process.

shift block: Dry sand, wood or cast iron blocks which fit prints on cope and drag pattern boards (Fig. S-9). The blocks are placed in the drag prints and the mold coped. The corresponding prints in the cope mold lower over the shift blocks which prevent a shift in the mold. The use of shift blocks is restored to only if the only flask available for the job is in no shape to trust. Shift blocks are also used to register cope and drag core molds which are not made in a flask. They are double-faced truncated pyramids or cones.

shifts: There are two classifications of shifts: *mold shift* and *core shift*.

A mold shift is when the parting lines are not matched when the mold is closed. The resulting is a casting offset or one that is mismatched at the parting. The causes are excessive rapping of a loose pattern, reversing the cope on the drag, too loose a fit or

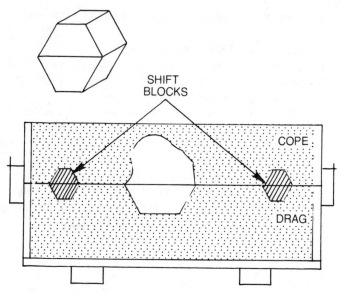

SHIFT
BLOCKS

COPE

DRAG

Fig. S-9. Shift block.

the pattern pins and dowels, faulty mismatched flasks, too much play between pins and guides, faulty clamping, improper fitting (racked) jackets and improper placing of jackets.

A core shift is caused by not aligning the halves of glued cores true and proper when assembling them.

shrink cavity and shrink depression: These defects are caused by lack of feed metal causing a depression on the concave surface of the casting. The shrink cavity is a cavity below the surface. It is not connected to the surface with a dendrite crystal structure.

shrink rules: Steel rules purchased from pattern supply houses that have the desired shrinkage for the particular use worked out over its length. Thus, if you purchase a 3/16 shrink rule (for brass), the rule will be divided into 12 inches but will actually be 12 inches and 3/16 inch long or 3/16 inch longer than a standard 12 inch rule. The shrink rule is used to dimension the pattern and will automatically compensate for the shrinkage chosen. Shrink rules can be purchased with any desired shrink or combinations needed including a double shrink.

silica flour: The most widely used inorganic additive is silica flour. It is used in steel sands to control and increase the hot, dry and green strengths. With the advent of the use of finer base sands for steel, silica flour is used much less now than in the past.

silicous clay: A clay which contains a high percent of silica.

sillimanite: A sand $Al_2 O_3 SiO_2$ used as a core sand for precision work. A very fine grained core can be made with sillimanite which will hold a dimension.

silicon brass and silicon bronze: An alloy containing over .5 percent silicon, 4 percent zinc, maximum and 98 percent copper, maximum.

Regular silicon bronze consists of 92 percent copper, 4 percent zinc and 4 percent silicon. It holds .302 pounds per cubic inch. A patternmaker's shrinkage is 3/16 inch per foot.

The pouring temperature for light castings is 2050 to 2250°F. For heavy castings, the temperature is 1900 to 2050°F.

No flux or deoxidizer is needed. It is contaminated easily with lead.

sil sand: The fine dust collected from the tops of sils and beams in the foundry. Used as a dry parting material in a shake bag or old sock.

sintering point: The temperature at which the molding material begins to adhere to the casting, or in the sintering test where the sand adheres to a platium ribbon under controlled conditions.

skeleton pattern: A frame work of wooden bars which represent the interior and exterior form as well as the metal thickness of the required casting. This type of pattern is only used for huge castings.

skin: The thin surface layer on a casting which is different from the main mass in composition and structure.

skin drying: Drying the surface of a mold by the direct application of heat from a torch or coke basket.

skull: The film of metal or dross remaining in a ladle after it has been emptied.

slab core: A flat plain core used in a mold to form a flat surface or as a cover core.

slag hole: The opening in the back of the cupola through which the slag is drawn from the molten metal. Also called the *slag notch*.

slag inclusion: Slag on the face of the casting and usually down the sides of the sprue. The cause is not skimming the ladle properly, not choking the sprue (keeping it brimming full from start to finish), sprue too big (cannot be kept choked) and gating system improperly choked.

slick and oval spoon: This tool is a must for all molders (Fig. S-10). Again the size needed is determined by the work involved. Most molders have at least four sizes from the little tiny one that is ¼ inch wide to the big one 2 inches wide. This tool is called a *doubler ender* in the trade. One end is a slick similar to a heart trowel blade but more oval shaped. The opposite end is spoon shaped. The outside or working surface is convex like the back of a spoon. Its inner face is concave. This face is never used and therefore is usually not finished smooth. When new it is painted black. Both faces of the slick blade are highly polished.

The double ender is a general use molding tool used for slicking flat or concave surfaces, open-up sprues, etc.

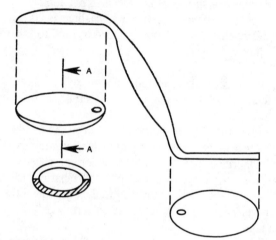

Fig. S-10. Slick and oval spoon.

slicking: Smoothing a section of a mold with the hands or a molder's slicker. Over slicking can result in blows, and scabs.

slip jacket: A metal or wood frame to place around a snap or slip flask mold after the flask has been removed during pouring (Fig. S-11).

slot sprue: A rectangular shaped sprue widely used in aluminum and magnesium casting (Fig. S-12).

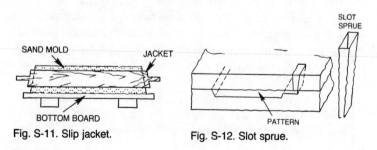

Fig. S-11. Slip jacket.

Fig. S-12. Slot sprue.

slush casting: A casting made from a low melting alloy which freezes over a wide range of temperatures. The molds are metal, usually brass or bronze, and mounted on trunnions. They are filled with metal with a hand ladle and rocked back and forth until an inner shell of metal is formed. The center, which is still liquid, is poured out leaving a hollow casting. Slip casting of metal is similar to slip casting a hollow wax form or clay form.

snagging: Rough grinding the gates, flash, etc., from a casting.

snap bands: Metal bands which are dropped into the cope and drag prior to the ramming of straight-sided snaps which reinforce the

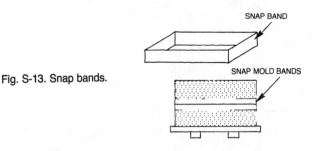

Fig. S-13. Snap bands.

mold at the parting line (cope and drag) to help prevent run outs (Fig. S-13). Also called *bands*.

snap flask: A snap flask is a flask, usually made of cherrywood. After the mold is made the flask can be removed by opening the flask and lifting it off the mold, leaving the mold as a block of sand on the bottom board (Fig. S-14).

Both cope and drag have a hinge in a corner *A*, and in corner *B*, a cam locking device. In operation the cope and the drag locks are closed tight and the mold made in the usual manner. When finished, the locks are opened and the flask is opened and removed from the mold.

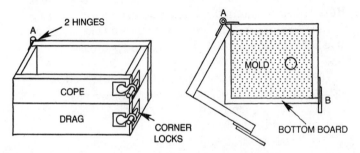

Fig. S-14. Snap flask. A. Hinges and B. cam locking devices.

The big advantage of the snap flask is that you need only one flask to make as many molds a day as you wish. With rigid flasks you need as many flasks as the number of molds you wish to put up at a time. I have seen small shops that had only three or four sizes of snaps and a variety of wooden floor flasks.

soda ash: Sodium carbonate used as a flux.

soldiers: Square wooden sticks dipped in a clay wash. They are used to reinforce and support the sand in the cope (Fig. S-15). Deep pockets of sand hang from the cope. They are placed in position during the ramming of the cope.

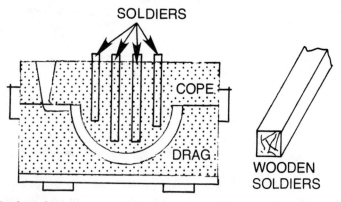

Fig. S-15. Solders.

solid contraction: The shrinkage a metal takes from its solidification point to room temperature.

solidification shrinkage: The shrinkage taken by a molten metal when going from a liquid to a solid state.

sorbite: When medium carbon steel is water quenched it results in a hard close-grained metal. If the quench is not as severe or slower, the steel will be softer. It is then called *sorbite or troostite*.

split pattern: A pattern that is made in two halves split along the parting line (Fig. S-16). The two halves are held in register by pins called *pattern dowels*. The pattern is split to facilitate molding.

The dowels hold the two halves of the pattern together in close accurate register, but at the same time are free enough that the two halves can be separated easily for molding, similar to the pins and guides of the flask.

The dowels are usually installed off center in such a manner that the pattern can only be put together correctly.

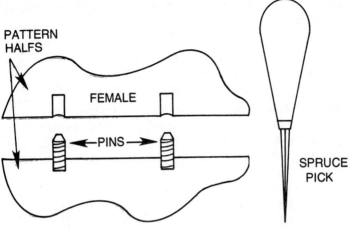

PATTERN
HALFS

FEMALE

←PINS→

SPRUCE
PICK

Fig. S-16. Split pattern.

Fig. S-17. Sprue stick.

sprue pick: A molder's tool used to draw a wooden set sprue from the cope (Fig. S-17). It consists of a double-ended tool having a sharp metal point on one end similar to a dart. The opposite end is pear shaped to give the tool the necessary driving weight. The point is driven into the top of the sprue with a flicking movement like throwing a dart. The sprue is simply lifted out. The tool provides the handle to do this. The sprue pick is also used to draw small wooden patterns from a mold.

stack core mold: A series of mold sections made in dry sand cores poured through a central sprue. Often rotated during pouring or centrifuging.

stainless steel: A wide range of iron alloys having a high content of chromium. They are highly resistant to oxidation and corrosion even at elevated temperatures. The original stainless produced in the U. S. in 1914 was 13.5 percent chromium and .35 carbon.

steadite: The eutectic of iron, iron phosphide and cementite.

steam gas porosity: This defect usually shows up as round holes similar to holes in Swiss cheese, just under the cope surface of the casting. It comes to light during machining. The cause is a wet ladle where the ladle lining was not properly and thoroughly dried.

In extreme cases the metal will kick and boil in the ladle. The practice of pigging the metal in a wet ladle and refilling it with the hopes that the pigged metal will finish drying the ladle is sheer folly.

step gate: A gate with one or more steps, used for two basic reasons (Fig. S-18).

- No. 1. To prevent a long metal fall into the mold which could cause mold erosion.
- No. 2. Used to produce a more elastic gating system to prevent hot tearing at its junction with the casting.

sticker: A lump or rat (bump) on the surface of the casting caused by a portion of the mold face sticking to the pattern and being removed with the pattern. This problem is caused by poor cleaning, shellacked, polished pattern, rough pattern, cheap

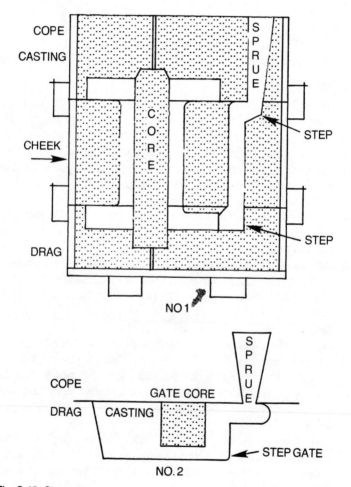

Fig. S-18. Step gate.

shellac, tacky shellac, sticky liquid parting, cold pattern against hot sand or an insufficient draft.

stock cores: Round cores of various diameters and length kept in the core room for those odds-and-ends type jobs with a missing round core box or kept to core holes through open sand castings (lifter plates, etc.). Stock cores are made from stock boxes or a coremaking machine resembling in looks and operation to a *sausage machine*.

These machines have tubes of various diameters which fit on the end (Fig. S-19). The sand is extruded through the tube by an auger turned by hand or motorized. A vent wire produces a vent

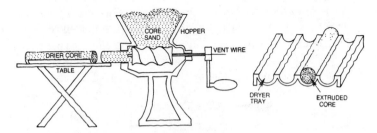

Fig. S-19. Stock cores.

through the core on its passage through the tube. The core is extruded on to corrugated driers for baking.

strike-off bar: Each time a mold is rammed up the sand must be struck off level with the flask cope and drag (Fig. S-20). The bar simply consists of a metal or hardwood straight edge of sufficient length.

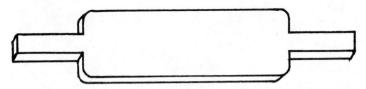

Fig. S-20. Strike-off bar.

stripping plate: A metal plate through which a pattern projects which conforms to the pattern very closely. It is used to mold straight cast tooth gears with no draft. The mold is rammed in the usual manner and the pattern stripped by pulling it down through the plate with a rack and pinion gear. The plate supports the sand between the teeth, allowing you to draw a pattern with 0 draft from the mold without damage to the mold.

sub-zero metal: Metals that will maintain the physical properties at sub-zero temperatures. Manganese bronze and aluminum bronze are among the sub-zero metals. They are not adversely affected at temperatures as low as minus 300°F.

sucker: The sucker is not a purchased tool but one made by the molder. It consists of two pieces of tubing and a tee of copper or iron (Fig. S-21).

It's use is to clean out deep pockets in molds where the bellows and lifter fail or if the pocket or slot is just too dirty to spend the time a lifter would take, or where it would be hard to see what you are doing with a lifter. The operation is very simple. You simply blow through the elbow with an air hose. This creates a vacuum in the long length which gives you in effect a vacuum cleaner with a long skinny snout. Now stick the long end down to where the problem is and blow through the elbow. These jobs will lift out small steel shot, a match stick or material which cannot be wetted such as parting powder or silica sand.

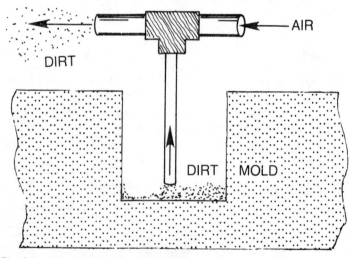

Fig. S-21. A sucker is made by the molder.

sulphur dome: An inverted dome placed over a pot or crucible of molten magnesium containing a high concentration of sulphur dioxide.

swab: A sponge, rag or hemp material used to dampen the sand around a pattern before drawing it.

sweep pattern: A sweep pattern consists of a board having a profile of the desired mold. When revolved around a suitable

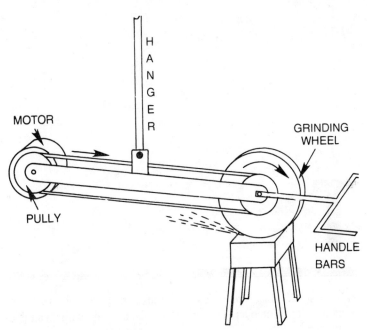

Fig. S-22. Swing frame grinder.

spindle or guide, it produces that mold. Two are usually required, one to sweep the cope profile and the other the drag profile.

swell: In this defect you have a casting which is deformed due to the pressure of the metal moving or displacing the sand. It is usually caused by a soft spot or a too soft mold.

swing frame grinder: A large grinder which is suspended from a crane or chain so that it can be maneuvered easily with a swinging action (Fig. S-22). Used to grind large castings which are too bulky or large to grind any other way.

synthetic molding sand: A molding sand compounded from selected individual materials which when mixed together produce the desired properties.

T

talc: Also called *soapstone*. It is a hydrous magnesium silicate which is used as a mold and core wash when combined with a suitable binder.

The most widely used homebrew mold and core wash is composed of one part molasses to 10 parts water with sufficient graphite added to produce the desired consistency.

tally mark: A mark or combination of marks to indicate the location of a loose piece of a pattern or core box. Sometimes called a *keeper mark*.

tamastone: A high strength gypsum used to make patterns, matchplates, core boxes, etc.

tap hole: The hole in the breast of a cupola or furnace through which the molten metal is tapped.

target boss: A boss on a casting from which all dimensions are taken or related to.

teapot spout ladle: A ladle designed like a teapot whereby the metal poured from the ladle comes from the bottom to insure clean slag free metal to the mold (Fig. T-1).

tellurium: An elementary metal, Te, obtained as a steel gray powder of 99 percent purity by the reduction of tellurium oxide or tellurite. It's chief use is as an alloy with lead to harden and toughen the lead.

tellurium bronze: Composed of 1 percent tellurium, 1.5 percent tin and the balance copper. It has an annealed strength of 40,000 PSI.

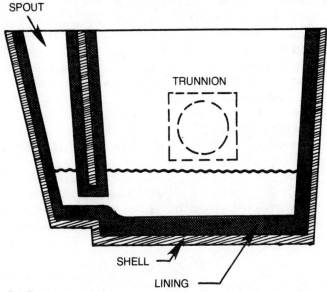

SPOUT

TRUNNION

SHELL ——

LINING ——

Fig. T-1. Teapot spout ladle.

Also used as a chill core wash. When a section of a core is washed with a tellurium wash. The area washed will act as a chill. A great tool to promote directional solidification on complex cored casting.

temper (verb): To mix water or other liquid with sand.

temper (noun): The moisture content of a tempered sand.

temper carbon: The free graphite (graphitic carbon) that precipitates from solution during the graphitizing of white cast iron.

templet: A thin piece of material with the edge contour made in the reverse of the surface to be formed or checked.

temporary pattern: A cheaply built pattern used to produce one or two castings.

tensile strength: Tensile strength is the force that holds the sand up in the cope. And, as molding sands are many times stronger in compression than tensile strength, we must take the tensile strength into account. Mold failure is more apt to occur under tensile forces.

Where compression strength is measured in pounds per square inch, the tensile strength of molding sands is measured in ounces per square inch.

The tensile strength which is the force required to pull the sample apart is determined very easily.

ternary alloy: An alloy that contains three principal elements. Gun metal of 88, 10 and 2 is a ternary alloy.

test bar: A standard specimen designed to permit determination of mechanical properties of a metal. Also called a *coupon*. It is poured from the same ladle of metal as the casting or is cast as an attachment to the casting and shipped with the casting.

Texas ramming: Term given to the method of ramming a flask cope and drag with the peen end of a molder's shovel and the back of the blade (Fig. T-2). The sand is *riddled* into the cope, filled and *peened* and then filled to overflowing and rammed by swatting with the back side of the shovel. Keep adding and swatting until firm, then strike off and finish in the usual manner. For shallow copes and drags with small, comparatively flat patterns, it can be quite fast. If executed correctly it produces good fast work, but it takes a lot of steam to keep it up for any length of time.

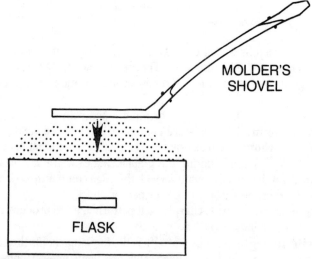

MOLDER'S SHOVEL

FLASK

Fig. T-2. Texas ramming.

threaded inserts: Threaded inserts are used where an accurate thread is necessary. They differ from screw shells in that they are machined from black bar stock and can be furnished in many different diameters and thread sizes (Fig. T-3). They produce a much better thread than the screw shells and can be designed to meet varying metal thicknesses surrounding the insert. Threaded inserts are used in chilled castings where drilling and tapping are impossible. Either an open-end or closed-end insert can be used.

Threaded inserts are used in a manner similar to that of screw shells. A hole is provided in the pattern large enough to allow the insert to draw freely and the insert is dropped into the hole before molding. If an open end insert is used, the insert fills with green sand when the mold is rammed and a nail can be pushed into the sand to reinforce it. Open end inserts can also be used by attaching them to cores.

The closed-end insert is rammed up in the same manner as the open-end type, but it is not necessary to fill it. A wooden plug pressed into the open end maintains the insert in its proper location in the mold. When the casting is poured the insert becomes part of it.

The outside surface of the insert is knurled to insure good fusion with the molten metal.

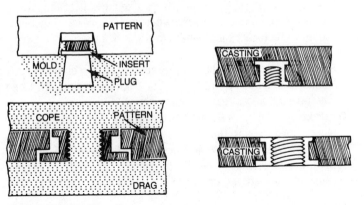

Fig. T-3. Threaded inserts.

tilting furnace: Any furnace that can be tilted to pour out the molten metal.

tin: A silvery white metal with a bluish tinge, melting point of 232°C and specific gravity of 7.298. It is widely used as a hot dip coating on steel (tin plate) or it is electroplated. A widely used metal as an alloy in a wide variety of casting metals and solders.

tin brass: A brass with over 6 percent tin and a zinc content even greater. An alloy seldom used in the foundry.

tin bronze: A bronze with 2 to 20 percent tin with the zinc less than the tin. The most commonly used foundry tin bronze is called straight bronze, 88-10-2 and gun metal. It is 88 percent copper, 10 percent tin and 2 percent zinc. Its weight per cubic inch is .315

pounds. Patternmaker's shrinkage is 3/16 inch per foot. Pouring temperatures for light castings is 2100 to 2300°F. For heavy castings it is 1920 to 2100°F.

tin tubes: Tin tubes find many diversified uses in foundry operations, but their principal use is for coring holes through heavy metal sections and reinforcing as well as ventilating intricate cores. They are also used as chills to initiate setting of heavier metal sections at the same rate as thin sections, thereby preventing shrinks and porosity.

These tubes made of tin plate have a free fit inside for the diameter listed. Tubes can also be furnished filled with core sand or made from perforated metal. Holes cored with plain tin tubes have a smooth surface after the sand is knocked out.

tramp iron: Iron shot, gates, sprues, nails, etc., in the molding sand due to improper conditioning of the sand and not using a magnetic separator to clean it. Tramp iron or brass can be the cause of blowing and chilling defects when they get rammed up in the sand next to the pattern when molding. Also they always seem to be located just where you go to cut a gate or punch a sprue hole.

transfer ladle: A ladle that may be supported on a crane bridge or monorail and used to transfer metal from the melting furnace to the pouring floors or a holding furnace.

triplexing: Making steel where the iron is melted in the cupola then transferred to the converter and blown. The blown heat is then transferred to the arc furnace for finishing and alloying.

trunnion: A boss-like projection (in pairs) on a flask, ladle, furnace, etc., used to lift and or rotate the object on an axis (Fig. T-4).

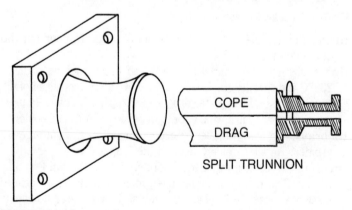

Fig. T-4. Trunnion.

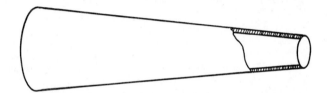

Fig. T-5. Tubular sprue cutter.

Flask trunnions are used for lifting the flask with a beam bail and to roll it over. Large slip and pop flasks are fitted with half trunnions or split trunnions—half on the cope and half on the drag. When the mold is closed it can be rolled on the trunnions.

tubular sprue cutters: A tapered steel or brass tube used to cut a sprue hole in the cope half of a sand mold (Fig. T-5). It is sold in sizes from ⅞ to 1¼ inches in diameter. All are 6 inches long.

tucking: Pressing sand with the fingers under flask bars, around gaggers and other places where the peen or rammer does not give the desired density. An important operation with hand ramming as well as some machine molded work.

tuyere: An opening in the cupola or blast furnace through which the blast air is forced.

U

upset: Any frame metal or wood used to increase the depth of the cope or drag (Fig. U-1). On a snap pop-off or slip flask it is usually screwed to the flask. In the case of rigid wood or metal flasks, they usually have tabs which extend down inside the flask being upset in place during molding and pouring. A square or round frame used to increase a sprue or riser height is also referred to as an upset.

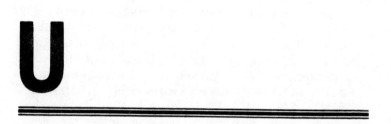

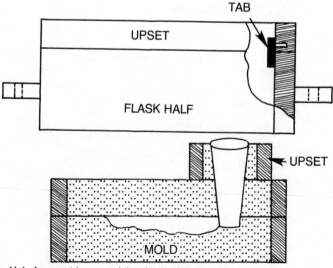

Fig. U-1. An upset increased the depth of the cope or drag.

U.S. sieve series: A series of nesting sieves used to determine the average grain finenesses and grain distribution of dry sands. The sieves are stacked in order from the most course on top to the finest on the bottom and a pan on the bottom. The sample to be checked is weighed and placed on the top sieve. The sieves are shook or vibrated. The sieves are dissassembled and the retained sand on each screen weighed. The percents are then plotted against the beginning known weight. A complete set consists of the following meshes: 4, 6, 8, 12, 16, 20, 30, 40, 50, 70, 100, 140, 200, 270 and the pan.

vanadium: An elementary metal V. A pale gray metal with a silvery luster with a specific gravity of 6.02 and a melting point of 3236°F. It will not oxidize in the air. In the percentages of .15 to 0.25 with a small quantity of chromium, it is widely used to alloy steel (*vanadium steel*).

ventilated flasks: A metal flask with round or oblong holes through its sides and ends. The holes allow the steam and gasses from pouring the casting to vent more freely than if the flask were not perforated. It is believed by some that the resistance of a solid flask wall to the escape of mold gasses builds up the pressure in the molds, leading to casting defects such as pin holes.

vent plate: A plate containing a series of vents (screen or slotted) used under an open-ended core box for blowing the cores (Fig. V-1).

vent plugs: A vent made in a permanent mold by drilling a hole in the mold and plugging it with a square plug (Fig. V-2). Vent screens and slotted vents installed in metal core blowing boxes are also called vent plugs.

vent wire: The vent wire is simply a slender pointed wire with a loop at its top. It is used to punch vent holes in the cope and drag of a sand mold to provide easy access of steam and gases to the outside during the pouring of the casting. The venting is done prior to the pattern removal. The vent wire is pushed down into

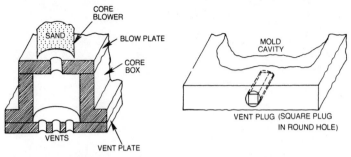

Fig. V-1. A vent plate.　　　　Fig. V-2. A vent plug.

the sand to within close proximity of the pattern. The first and second finger straddles the wire each time it is withdrawn as a guide.

vitrification point: The temperature upon heating at which clays reach the condition of maximum density and shrinkage (vitrified).

volumetric shrinkage: The volumetric shrinkage of a molten metal from its liquid state to its solidification temperature.

wash: Any number of refractory compounds used to coat molds, cores, ladles, cupola spouts, etc. Increases the refractiveness of the surface to which it is applied.

washburn core: A washburn core consists of a riser necking core which conforms to the shape of the casting at the point of riser attachment (Fig. W-1). Its purpose is to keep the riser open so that it can feed the casting and at the same time reduce the size of the riser attachment, resulting in a great saving in riser removal and grinding. The core can be made of various kinds of refractory

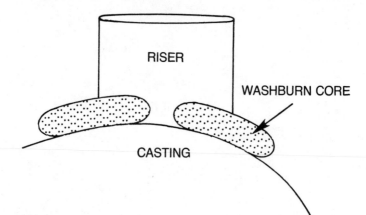

Fig. W-1. Washburn core.

material (fired ceramic shape) or a rodded oil sand core. There is a straight line relationship between the thickness of the core and the neck diameter.

washing and erosion: The sand is eroded and washed around in the mold. Some of it finds its way to the cope surface of the casting as dirt sand inclusions. It can come from the gating system or in the mold cavity. The causes include a too low hot strength, a too dry molding sand, poor gating design, a deep drop into the mold, washing at the point of impact, metal washing over a sharp edge at the gate, metal hitting against a core or a vertical wall during the pouring.

water soluble wax: A wax primarily used in investment casting wax. Patterns work as spacers between the blades of high speed impeller patterns. The pattern is assembled with regular wax and water soluble wax spacers. When complete the water soluble spacers are dissolved leaving precise spacing between the blades.

web: A plate or member lying between heavier sections (Fig. W-2).

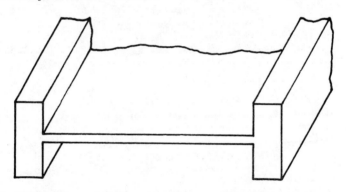

Fig. W-2. Web.

weep hole: A hole placed in a casting to allow drainage of moisture.

well board: A board fitted with pins or guides to match a flask (Fig. W-3). The board has a well into which a pattern board fits flush. The use of well boards allows a quick pattern change and standardizes the pattern mounts.

whirl gate: A gate or sprue arranged in a manner to introduce the metal tangentially, thereby imparting a swirling motion (Fig. W-4).

whistler: A small (¼inch) opening through the cope of a mold at the highest point on the casting used to tell when the mold is full when pouring (Fig. W-5).

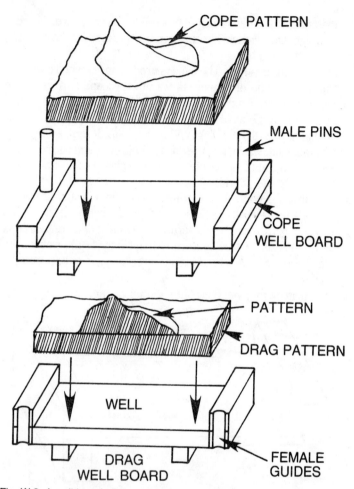

COPE PATTERN

MALE PINS

COPE
WELL BOARD

PATTERN

DRAG PATTERN

WELL

DRAG
WELL BOARD

FEMALE
GUIDES

Fig. W-3. A well board allows quick pattern changes.

white iron: A cast iron in which all the carbon is in the combined state and none is thrown out during solidification as free graphite. It is the chemical compound of most of the carbon with iron as opposed to gray iron. Most of the carbon is in the form of graphite, mechanically mixed with the iron. White iron is used to cast parts where high wearing qualities are desired. It is extremely brittle and cannot be machined by normal methods.

wind box: The chamber surrounding a cupola through which the blast air is conducted to the tuyeres. The wind box has a pressure equalizing effect, therefore each tuyere will receive approximately the same volume and pressure.

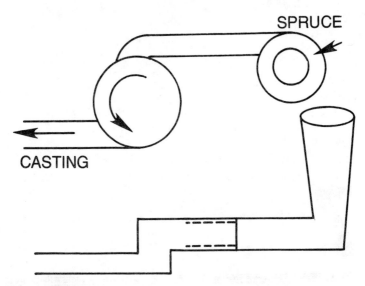

Fig. W-4. Whirl gate.

wood flour: All wood flours are not the same. Foundry wood flours should be a wood floor with the largest portion of resinous material removed. They should possess a low ash content. The normal percentages used in natural and synthetic sand mixes usually run up to 1½ percent. In core mixes use ¼ to ½ percent cereal and ½ percent wood flour. Wood flour additions will reduce volume changes, hot strength permeability and dry strength while it increases green compression strength, mold hardness, moisture, deformation and density.

wood's metal: 25 percent lead, 12.5 percent tin, 50 percent bismuth and 12.5 percent cadmium. This alloy has a melting temperature of 154.4°F.

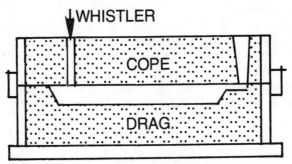

Fig. W-5. A whistler is a small opening through the cope of a mold.

Z

zinc: A bluish white crystalline metal which melts at 419°C and boils at 907°C. Zinc is a widely used metal for alloying. Also used in various chemical reactions, plating, etc.

zinc tracks: This defect is found on the cope surface of high zinc alloy castings. The defects are caused by the zinc distilling out of the metal during pouring. This zinc oxide floats up to the cope and form worm track lines on the casting when the metal sets against the cope. The problem is caused by pouring too hot (metal flaring) in ladle or crucible or pouring the mold too slow with insufficient gates. The mold must be filled quickly before the damage can be done.

zircon sand: Zirconium silicate comes chiefly from beach sands. It is used for mold facings, cores and when flour is used as mold and core washes.

zirconium: A silvery white metal which melts at approximately 1850°C and has the specific gravity of 6.5. It is difficult to reduce to a metal form as it combines easily with oxygen, nitrogen, carbon and silicon. It is produced by reacting *zircon sand* with carbon. Used as an alloy with ferrous and nonferrous metals and widely used as powerful metal deoxidizer.

Part 3
Metal Casting Statistical Data

Weights and Measures

60 seconds 1minute
90 degrees ....1 quadrant
60 minutes..1 degree
4 quadrants or 360 degrees ....1 circle
30 degrees..1 sign

Units of Measure

Acre = 208.71 feet square = 43,560 square feet = 4,480 square yards = 0.40687.

Hectares = 4,046.87 square meters.

Barrel = 196 pounds (Flour = 42 Gal. Oil (Standard Oil Co.).

Board Foot = one square foot, one inch thick.

Bushel = 4 pecks = 32 quarts = 2,150.42 cubic inches = 1.24446 cubic feet = 35.23928 liters.

Cable (Cable length) = 720 feet = 120 fathoms = 219.457 meters.

Chain = 100 feet = 100 links = 30.48 meters.

Dram (apothecary) = 3 scruples = 60 grains = 3.888 grammes.

Fathom = 6 feet = 1.829 meters.

Foot = 12 inches.

Furlong = 660 feet = 40 rods, perches or poles = ⅛ mile = 201.17 meters.
Gallon = 231 cubic inches = 3.78543 liters = 3,785.43 cubic centimeters.
Gill = ¼ pint.
Grain = 0.0648 grammes = 64.8 milligrammes.
Hogshead = 63 gallons = 2 barrels (31.5 gallons capacity) = 238.48 liters.
Inch = 2.54 centimeters = 25.4 millimeters.
Karat = 200 milligrammes = 0.2 grammes = 3.0865 grains.
Kilogramme = 1,000 grammes = 2.20462 pounds avd.
Kilometer = 1,000 meters = 3,280.83 feet = 0.62137 miles.
Knot (Nautical or geographical miles) = 6,080.2 feet = 1.15155 miles = 1.85325 kilometers = 1 minute of earth's circumference.
League = 15,840 feet = 3 miles = 4.828 kilometers.
Link = one hundreth of measuring chain = 12 inches (Engineer's chain) = 7.92 inches (Surveyor's chain) = 20 centimeters (Metric Chain).
Liter = 1,000 cubic centimeters = 61.023 cubic inches = 0.0353 cubic feet = 2.1134 liquid pints = 0.2642 gallons.
Meter = 39.37 inches = 3.28 feet.
Miles = 5,280 feet = 1,760 yards. A square mile equals 640 acres = 2.59 square kilometers.
Milligram = 0.001 grammes = 0.015432 grains.
Millimeter = 0.001 meters = 0.03937 inches.
Ounce, Apothecary. Same as troy ounce = 480 grains = 31.104 grammes. Avoirdupois = 437.5 grains = 28.35 grammes = 0.9115 ounce troy or apothecary. Troy (for gold and silver) = 480 grains = 20 pennyweight = 31.104 grammes = 1.097 ounces avd.
Peck = 0.25 bushels = 8.81 liters.
Pennyweight = 24 grains = 1.555 grammes.
Pint, liquid = 0.125 gallons = 0.4732 liters. Dry = 0.5 quarts = 0.5506 liters.
Pipe or Butt = 126 gallons = 2 hogsheads = 476.96 liters.
Pounds, Avoirdupois = 7,000 grains = 16 ounces (adv.) = 0.4536 kilogrammes.
Troy or Apothecary = 5,760 grains = 12 ounces = 0.3732 kilogrammes.
Quart, liquid = 0.25 gallons = 0.94634 liters. Dry = 0.03125 bushels = 67.2 cubic inches = 1.1 liters.
Rod or Perch or Pole = 16.5 feet = 5.5 yards = 5.0292 meters.
Rood = 0.25 acres = 40 square rods = 1,210 square yards = 1,011.72 square meters.
Scruple = 20 grains = 1.296 grammes.
Section of land = 1 mile square = 640 acres.
Stone = 14 pounds (avd) = 6.35 kilogrammes.
Ton (gross) Displacement of water = 35.88 cubic feet = 1,016 cubic meters. (gross or long) = 2,240 pounds (avd.) = 1.12 short or net tons = 1,016.05 kilogrammes = 1.01605 metric tons. (net or short) = 2,000 pounds (avd.) = 20 hundredweight = 907.185 kilogrammes = 0.907185 metric tons = 0.892857 long tons. (metric) = 2,204.62 pounds (avd.) = 1.10231 net tons = 0.9842 long tons = 1,000 kilogrammes.
Cubic Yard = 27 cubic feet = 46,656 cubic inches = 0.76456 cubic meters. Square yard = 9 square feet = 1,296 square inches = 0.836 square meters. Yard = 3 feet = 36 inches = 0.9144 meters.

Physical Properties of Elements

	Symbol	Atomic Weight	Density Pounds/Cubic Inches 20°C	Approximate Melting Points Centigrade	Fahrenheit
Aluminum	Al	26.98	0.0975	660	1220
Antimony	Sb	121.76	0.2391	630	1167
Arsenic	As	74.91	0.2070	Sublimes	Sublimes
Beryllium	Be	9.013	0.067	1277	2332
Bismuth	Bi	209.00	0.3541	271	520
Boron	B	10.82	0.085	2300	4172
Cadmium	Cd	112.41	0.313	321	610
Calcium	Ca	40.08	0.0560	838	1540
Chromium	Cr	52.01	0.260	1875	3407
Cobalt	Co	58.94	0.322	1495	2723
Copper	Cu	63.54	0.324	1083	1981
Gold	Au	197.0	0.698	1063	1945
Indium	In	114.82	0.264	156	313
Iron	Fe	55.85	0.284	1536	2797
Lead	Pb	207.21	0.41	327	621
Lithium	Li	6.94	0.0190	180.54	356.97
Magnesium	Mg	24.32	0.0628	650	1202
Manganese	Mn	54.94	0.270	1245	2273
Mercury	Hg	200.61	0.4894	−38.36	−37.05
Molybdenum	Mo	95.95	0.369	2610	4730
Nickel	Ni	58.71	0.322	1453	2647
Phosphorous (yellow)	P	30.95	0.06	44.1*	111.4
Platinum	Pt	195.09	0.775	1773	3217
Potassium	K	39.1	0.031	63.7	146.7
Rhodium	Rh	102.91	0.447	1966	3571
Selenium	Se	78.96	0.174	217	423
Silicon	Si	28.06	0.084	1410	2570
Silver	Ag	107.88	0.379	961	1761
Sodium	Na	22.99	0.035	98	208
Sulfur	S	32.06	0.0748	119	246
Tellurium	Te	127.61	0.225	450	840
Tin	Sn	118.70	0.264	232	450
Titanium	Ti	47.90	0.163	1668	3035
Tungsten	W	183.86	0.697	3410	6170
Vanadium	V	50.95	0.220	1900	3450
Zinc	Zn	65.38	0.258	419	787
Zirconium	Zr	91.22	0.234	1852	3366

*Ignites in air 34°C

Characteristics of Castings
of Various Metals and Sizes

Light Gray Iron
Fineness...........................175
Clay 12%
Moisture.........................7.4%
Permeability.......................15
Green Compression..........4.0

Medium Gray Iron
Fineness...........................111
Clay 15%
Moisture..........................7.5%
Permeability.......................40
Green Compression..........4.0

Heavy Gray Iron
Fineness.............................73
Clay 18%
Moisture.........................7.6%
Permeability.......................70
Green Compression 5.0

Heavy Brass
Fineness...........................108
Clay 12%
Moisture............................7%
Permeability.......................51
Green Compression 4.0

Light to Medium Brass
Fineness...........................218
Clay 13%
Moisture............................8%
Permeability.......................18
Green Compression 4.0

Aluminum
Fineness...........................232
Clay 19%
Moisture............................8%
Permeability.......................18
Green Compression 5.0

Notice the two differences between the sands—the grain fineness and the permeability required in a sand used to make a mold for a gray iron casting compared to the fineness and permeability required in a sand for use in making a mold for aluminum.

Popular Additives

Pitch ....1.3 pound per quart
Corn Flour............1.2 pound per quart
Linseed Oil....... ..1.9 pound per quart
Bentonite ....1.5 pound per quart
Silica Flour ....2.2 pound per quart
Wood Flour.........10 ounces per quart

Sea Coal ....1.5 pound per quart
Goulac...............2.5 pound per quart
Gilsonite ....1.3 pound per quart
Fire Clay......... .2.0 pound per quart
Silica Sand.... .3.0 pound per quart

Conversion Tables

Volume

Multiply		By	To Obtain
Cubic Centimeters			
Dry Volume	(cm.^{3}or cu.cm.)	0.061023	Cubic Inches
Liquid Volume	(c.c.)	.001000	Liters
Liquid Volume	(c.c.)	0.033814	U.S. Fluid Ounces
Cubic Meters	(m.^{3}or cu. m.)	264.17	Gallons
Cubic Meters	(m.^{3}or cu. m.)	61,023.	Cubic Inches
Cubic Meters	(m.^{3}or cu. m.)	35.315	Cubic Feet
Cubic Meters	(m.^{3}or cu. m.)	1.3079	Cubic Yards
Cubic Feet		1,728.	Cubic Inches
Cubic Feet	(ft.^{3}or cu. ft.)	28,317.	Cubic Centimeters
Cubic Feet	(ft.^{3}or cu. ft.)	0.028317	Cubic Meters
Cubic Feet	(ft.^{3}or cu. ft.)	28.32	Liters
Cubic Feet	(ft.^{3}or cu. ft.)	0.037037	Cubic Yards
Cubic Feet	(ft.^{3}or cu. ft.)	7.4805	U.S. Gallons
Cubic Yards	(ft.^{3}or cu. ft.)	27.	Cubic Feet
Cubic Yards	(yd.^{3}or cu. yds.)	0.76456	Cubic Meters
Liters	(yd.^{3}or cu. yds.)	61.02	Cubic Inches
Liters		0.03531	Cubic Feet

Mass and Weight

Multiply	By	To Obtain
Grams (g.)	0.035274	Ounces Avoirdupois
Grams (g.)	0.0022046	Pounds Avoirdupois
Kilograms (kg.)	2.2046	Pounds Avoirdupois
Ounces Avoirdupois (oz.av.)	28.350	Grams
Ounces Apothecary or Troy (oz.ap.or t.)	31.103	Grams
Pounds Avoirdupois (lb.av.)	16	Ounces Avoirdupois
Pounds Avoirdupois (lb.av.)	453.59	Grams
Pounds Avoirdupois (lb.av.)	0.45359	Kilograms
Pounds Apothecary or Troy (lb.ap.or t.)	12.	Ounces Apothecary or Troy
Pounds Apothecary or Troy (lb.ap.or t.)	373.24	Grams
Metric Tons (t)	1,000.	Kilograms
Metric Tons (t)	2,204.6	Pounds Avoirdupois
Metric Tons (t)	1.1023	Short Tons
Short Tons	2,000.	Pounds Avoirdupois
Short Tons	907.18	Kilograms
Long Tons	2,240.	Pounds Avoirdupois
Long Tons	1,016.0	Kilograms
Assay Tons	29.167	Grams

Pressure

Multiply	By	To Obtain
Kilograms per Square Centimeter	14.22	Pounds per Square Inch
Kilograms per Square Centimeter	1.024	Short Tons per Square Foot
Kilograms per Square Centimeter	0.9678	Atmospheres
Pounds per Square Inch	0.07031	Kilograms per Square Centimeter
Pounds per Square Inch	0.0720	Short Tons per Square Foot
Pounds per Square Inch	0.06804	Atmospheres
Pounds per Square Inch	2.307	Feet of Water at 39.2°F.
Pounds per Square Inch	2.036	Inches of Mercury at 0°C
Pounds per Square Foot	0.4882	Grams per Square Centimeter
Pounds per Square Foot	0.00050	Short Tons per Square Foot

Length

Multiply		By	To Obtain
Millimeters	(mm.)	0.001	Meters
Millimeters	(mm.)	0.039370	Inches
Centimeters	(cm.)	0.01	Meters
Centimeters	(cm.)	0.39370	Inches
Decimeters	(dm.)	0.1	Meters
Decimeters	(dm.)	3.9370	Inches
Meters	(m.)	39.370	Inches
Meters	(m.)	3.2808	Feet
Kilometers	(km.)	1,000.	Meters
Kilometers	(km.)	3,280.8	Feet
Inches	(in.)	25.400	Millimeters
Inches	(in.)	2.5400	Centimeters
Feet	(ft.)	12.	Inches
Feet	(ft.)	30.480	Centimeters
Feet	(ft.)	0.30480	Meters
Yards	(yd.)	91.440	Centimeters
Yards	(yd.)	0.91440	Meters
Statute Miles	(st. mi.)	5,280.	Feet
Statute Miles	(st. mi.)	1,760	Yards

Domestic Weights and Measures

Avoirdupois Weight

437½ grains	1 ounce
16 ounces	1 pound
25 pounds	1 quarter
4 quarters	1 cwt.
20 cwt.	1 ton
2240 pounds	1 long ton

Troy Weight

24 grains	1 pennyweight
20 pwt.	1 ounce
12 ounces	1 pound

Apothecaries' Weight

20 grains	1 scruple
3 scruples	1 dram
8 drams	1 ounce
12 ounces	1 pound

Dry Measure

2 pints	
8 quarts	
4 pecks	
36 bushels	

Liquid Measure

4 gills	1 pint
2 pints	1 quart
4 quarts	1 gallon
31½ gallons	1 barrel
2 barrels	1 hogshead

Linear Measure

12 inches	1 foot
3 feet	1 yard
5½ yards-16½ feet	1 rod
320 rods-5280 feet	1 statute mile
6080.20 feet	1 nautical mile

Cubic or Solid Measure

1728 cu. inches	1 cu. foot
27 cu. feet	1 cu. yard
128 cu. feet	1 cord
40 cu. feet	1 ton of ship cargo

Metric Weights and Measures

Metric weights and measures form a decimal system based upon the meter. For convenience, the liter is used as the unit of capacity and the gram as the unit of weight.

The liter equals one cubic decimeter.

The gram is the weight of one cubic centimeter of water at its greatest density.

Parts and multiples of the unit are indicated by the following prefixes·

Milli	(m) meaning	1/1000
Centi	(c) meaning	1/100
Deci	(d) meaning	1/10
Deka	(dk) meaning	10
Hecto	(H) meaning	100
Kilo	(K) meaning	1,000
Myria		10,000

Surface Measure

144 sq. inches	1 sq. ft.
9 sq. feet	1 sq. yard
30¼ sq. yds.	1 sq. rod
160 sq. rods	1 acre
640 acres	1 sq. mile
1 acre	43,560 sq. ft.

Comparisons

U. S. Bushel	2150.42 cu. inches
Br. Imp. bushel	2218.2 cu. inches
U. S. gallon	231 cu. inches
7.481 U. S. gallons	1 cu. foot
6.229 Br. Imp. gallons	1 cu. foot
6 U. S. gallons	5 Br. Imp. gallons
1 cord	about 103 bushels
1 meter	39.37 in. (U. S. statute)
1 liter	61.022 cu. in. (U. S. Statute)
1 gram	15.42 grains (U. S. statute)
25.4 mm.	1 inch
30.48 cm.	1 foot
1 meter	3.281 feet
1.6093 kilometer	1 mile
6.4515 sq. cm.	1 sq. inch
1 sq. meter	10.764 sq. ft.

Comparisons

1 sq. meter	1.550 sq. inches
1 cu. meter	264.4 U. S. gallons
1 kilogram	2.2046 pounds
1,000 kilograms	1 metric ton
1 kg. per sq. cm.	14.223 lbs. per sq. inch

Area

Multiply		By	To Obtain
Square Millimeters	(mm.² or sq.mm.)	0.0015500	Square Inches
Square Centimeters	(cm.²or sq.cm.)	0.15500	Square Inches
Square Meters	(M.² or sq. m.)	1,000.	Square Centimeters
Square Meters	(m.²or sq. m.)	10.764	Square Feet
Square Inch	(in.²or sq.in.)	645.16	Square Millimeters
Square Inch	(in.²or sq.in.)	6.4516	Square Centimeters
Square Feet	(ft.²or sq.ft.)	144.	Square Inches
Square Feet	(ft.²or sq.ft.)	929.03	Square Centimeters
Square Feet	(ft.²or sq.ft.)	0.092903	Square Meters
Square Yards	(yd.²or sq.yd.)	0.83613	Square Meters
Acres	(A)	43,560.	Square Feet
Acres	(A)	4,840.	Square Yards
Acres	(A)	4,046.9	Square Meters
Square Miles	(mi.²or sq.mi.)	640.	Acres

Shrinkage of Castings Per Foot

Metals	Fractions of an Inch
Pure Aluminum	13/64
Nickel Aluminum Casting Alloy	3/16
Special Casting Alloy, made by the Pittsburgh Reduction Co.	11/64
Iron, small Cylinders	1/16
Iron, Pipes	⅛
Iron, Girders, Beams, Etc.	1/64
Iron, Large Cylinders, Contraction of Diameter at Top	⅝
Iron, Large Cylinders, Contraction of Diameter at Bottom	5/64
Iron, Large Cylinders, Contraction in Length	3/32
Cast Iron	⅛
Steel	¼
Malleable Iron	⅛
Tin	1/12
Britannia	1/32
Thin Brass Castings	11/64
Thick Brass Castings	5/32
Zinc	5/16
Lead	5/16
Copper	3/16
Bismuth	5/32

Weight of Various Materials

Material	Average Per Cubic Feet (Pounds)
Lime	
Quick, loose lumps	53
Quick, fine	75
Stone, large rocks	168
Stone, irregular lumps	96
Masonry	
Granite or Limestone	165
Mortar, rubble	154
Dry	138
Sandstone, dressed	144
Oils	
Engine	55
Crude	48
Petroleum	55
Gasoline	43
Rock	
Chalk	145
Granite	165
Gypsum	143
Rock	
Sandstone	144
Pumice Stone	57
Quartz	165
Salt, coarse	45
Salt, fine	49
Shales	162
Slate, American	175
Brick	
Common red	100
Fire Clay	150
Silica	128
Chrome	175
Magnesia as brick or fused in furnace	160
Dead Burn Grain-Heavy Type	
Magnesite as shipped	112
Cement	
Portland	94
Hydraulic	60

Material	Average Per Cubic Feet (Pounds)
Cork	15
Coal and Coke	
Anthracite	55-66
Bituminous	50-55
Charcoal	18.5
Coke	31
Concrete	
Cement, fine	137
Rubble, coarse	119
Earth	
Loam, dry, loose	76
Loam, packed	95
Loam, soft, loose mud	108
Loam, dense mud	125
Sand	
Dry and loose	100
Dry and packed	110
Wet and packed	130
Gravel, packed	118
Water	
Water as ice	58.7
Water at 32 degrees Fahrenheit	62.4
Water at 212 degrees Fahrenheit	59.6
Woods, Dry	
Apple	48
Beech	49
Birch	40
Cedar, American	32
Chestnut	41
Ebony	76
Elm	35
Hemlock	25
Hickory	42
Ironwood	114
Mahogany	41 to 53
Maple	43
Oak, live	47
Oak, white	47
Pine, white	27
Pine, yellow northern	34
Pine, yellow southern	45
Spruce	37
Walnut	42

Sand Mixes

Half & Half:

Brass sand for general all around work.
50 percent Albany 00 or similar.
50 percent 120 mesh sharp silica.
4.5 percent Southern bentonite.
4.5 percent Moisture
2 percent Wood flour 200 mesh.

72 Hour Cement Sand:

Mold first day, wash second day, pour third day.
10 parts silica sand.
1 part Hi-Early Cement.
1 percent by weight 200 mesh wood flour.
Oil patterns to obtain a parting when molding.
On the 72 hour cement sand the reference to wash the second day refers to a core wash. This type of mold would be considered a core sand mold.

Synthetic Mix for Light Brass

160 mesh wash float penn silica
2 percent wood flour

4.5 percent southern bentonite'
4.5 percent moisture

Synthetic Mix for Heavy & Medium Brass

Nevada 120 silica
2% 200 mesh wood flour

4.5% southern bentonite
4.5% moisture

Rebound Facing:

3 parts air float sea coal
1 part Goulac
10 parts molding sand.

Synthetic Mix For Brass & Bronze

AFS fineness 60 to 80
95.5 percent by weight dry sharp silica
4.0 percent 50/50 Southern-Western Bentonite

(98.5 percent silica content)
0.5 percent corn flour
1.5 percent Dextrine

Dry Sand Mix for Brass & Bronze

Use system sand tempered with glutrin water, 1 pint of glutrin to 5 gallons of water. Bake till dry at 350°F (5 percent pitch can be added to the facing sand).

Dry Sand Molds for Phosphor Bronze

System sand such as Albany or a synthetic sand with the addition of one part wheat flour (bakery sweepings) to each 40 parts molding sand. Temper with clay water 30° Baume (bonding clay, red clay or blue clay, not fire clay). For aluminum bronze leave out flour if desired.

Cement Sand Mix for Large Brass & Bronze

6.5 percent moisture   12 percent Portland cement.
81.5 percent sharp silica sand approximately AFS 40 fineness.

Facing for Plaques & Art Work

5 shovels fine natural bonded sand (Albany or equal).
¼ shovel powdered sulphur   ¼ shovel iron oxide.
Spray mold face with molasses water no later than 10 seconds after drawing pattern
(this is one of the most important steps in producing a plaque). The molasses water in
this case consists of one part blackstrap molasses to 15 parts of water by volume.
Allow the mold to air dry a bit before pouring. Brass sand can be made up synthetically
or partly synthetic and partly natural sand by choosing the correct base sand for the
class of work and bonding with 50:50 south and west bentonite. A very popular all
around mix is the brass mix.

Brass Mix:

90 pounds AFS-90-140 grain silica
10 pounds naturally bonded sand 150-200 fine.
4 pound southern bentonite
1 pound wood flour.
       To make any nonferrous mold suitable for skin dry work or dry sand work, simply
temper facing with glutrin water or add up to 1 percent pitch to the facing.

Hi Nickel Facing Sand:

15 parts silica AFS G. F. 120 and one-half part bentonite.
3 parts system sand and one-half part fire clay.
       For dry sand molds a good all around sand for brass and bronze is a brass and
bronze mix.

Brass & Bronze Mix:

Dry new sand, 95.5 pounds
Bentonite, 3 pounds
Corn flour, 0.2 pounds
Dextrine, 1.3 pounds
       In general for aluminum you can use a much finer base sand, synthetic or
naturally bonded sand, due to the low melting and pouring range of aluminum. As a
guide line, the two sands given here would cover 90 percent of all aluminum casting
work.

Aluminum Mix for Naturally Bonded Sand

Grain fineness, 210 to 260
Clay content, 15 to 22 percent
Moisture 6 percent
This should give a green strength of 6 to 8 and a permeability of 5 to 15.

Aluminum Mix for Synthetic Sand

Grain fineness (washed & dried silica), 70 to 160
Southern bentonite, 4 to 10 percent
Moisture, 4 to 5 percent
       When you compare these two basic molding sands you will note that the synthetic
sand is more open with a resulting higher permeability with a lower moisture content.
This sand, in areas of low humidity, is hard to keep wet enough to mold with due to
evaporation of the already low moisture required. If you lose say 2 percent of the
moisture on a sand carrying 4 percent you dropped 50 percent. Lose 2 percent of your
6 percent moisture with a naturally bonded sand and you have dropped only 33-⅓
percent which will not materially affect its moldability.

Albany Molding Sand

No. 00 AFA classification No. 1E (AFA-American Foundryman's Association)
Use: Very small iron, brass and aluminum.

No. 0 AFA classification No. 2-D or E.
Use: Medium weight castings where a smooth surface is essential.

No. 1AFA classification No. 3-D.
Use: Medium weight iron and brass general bench work.

No. 1½ AFA classification No. 3-D
Use: Medium to chunky iron & brass.

No. 2 AFA classification No. 4-D
Use: Medium iron-heavy brass.

No. 2½ AFA classification No. 4-D.
Use: Side floor molding and medium heavy iron & brass.

No. 3 AFA classification No. 5-D.
Use: Heavy iron and dry sand work.

Cupola Charges

Some typical charges for various castings are given below. The symbols used to denote the metals are:

Si—Silicon
S—Sulphur
Mn—Manganese
P—Phosphor
Tc—Total carbon

Product & Analysis   *Charge*

Pressure Castings

Tc—3.25%	Pig Iron 50%
Si—1.25%	Steel Scrap 25%
Mn—0.65%	Domestic Scrap 25%
P—0.20%	
S—0.10%	

General Castings (Soft)

Tc—3.40%	Pig Iron 50%
Si—2.60%	Domestic scrap 45%
Mn—0.65%	Steel scrap 5%
P—0.30%	
S—0.10%	

Plow Shares

Tc—3.60%	Pig iron 45%
Si—1.25%	Steel 15%
Mn—0.55%	Domestic Scrap 40%
P—0.40%	
S—0.10%	

The difference in silicon content between piston rings & plow shares

Product & Analysis   *Charge*

Valves & Fittings

Tc—3.30%	Pig Iron 50%
Si—2.00%	Steel Scrap 10%
Mn—0.50%	Purchased scrap 40%
P—0.60%	
S—0.10%	

Piston Rings

Tc—3.50%	Pig Iron 60%
Si—2.90%	Scrap or Sprues 40%
Mn—0.65%	
P—0.50%	
S—0.06%	

Also a very good mix for light miscellaneous castings.

Machinery Iron #1(thin section not over 1 inch thick)

Tc—3.25%	Pig Iron 50%
Si—2.25%	Scrap 50%
Mn—0.50%	
P—0.50%	
S—0.09%	

Product & Analysis	*Charge*	*Product & Analysis*	*Charge*

Machinery Iron#2 (section 1 to 1½ inches thick)

Tc—3.25%
Si—1.75%
Mn—0.50%
P—0.50%
S—0.10%

Pig Iron 50%
Steel 10%
Scrap 40%

Machinery Iron #3 (section 2 to 3 inches thick)

Tc—3.25%
Si—1.25%
Mn—0.50%
P—0.50%
S—0.10%

Pig Iron 50%
Steel 25%
Scrap 25%

High Strength Cupola Iron

Tc—2.75%
Si—2.25%
Mn—0.80%
S—0.09%

Steel Scrap 85%
Returns (sprues, gates, etc.) 10%
Ferro silicon 5%

Note: You will note that in machinery iron 1, 2 and 3, we called for the same Mn—0.50 and P—0.50, but note the difference in silicon from 1 to 3. Because of the cooling rate difference for the section thickness, silicon promotes free carbon as does slow cooling, thus the adjustment in silicon downward as we get into a slower cooling rate for thick sections.

Composition of Gray Iron Castings

Small Wheels

Si—1.75 to 2.25
S—max. 0.08
P—0.40 to 0.60
Mn—0.50 to 0.70

Automobile Castings

Si—1.75 to 2.25
S—max .08
P—0.40 to 0.60
Mn—0.60 to 0.80

Chilled Castings

Si—0.75 to 1.25
S—Max 0.10
P—0.20 to 0.40
Mn—0.80 to 1.20

Agriculture Castings

Si—2 to 2.50
S—max 0.09
P—0.60 to 0.80
Mn—0.50 to 0.80

Permanent Molds

Si—2 to 2.25
S—max 0.07
P—0.20 to 0.40
Mn—0.60 to 0.90

Collars & Couplings

Si—1.75 to 2
S—max 0.08
P—0.40 to 0.50
Mn—0.50 to 0.70

Brake Shoes

Si—1.40
S—max .10
P—0.50
Mn—0.50 to 0.70

Ornamental Iron

Si—2.25 to 2.70
S—max 0.09
P—0.7 to 1
Mn—0.40 to 0.60

Wood Working Machinery

Si—1.75 to 2.25
S—max 0.09
P—0.50 to 0.70
Mn—0.50 to 0.70

Ball Mill Grinding Balls

Si—0.90 to 1.20
S—max 0.15
P—0.20
Mn—0.60 to 1(Tc low)

Small Pulleys

Si—2.25 to 2.75
S—max 0.08
P—0.60 to 0.90
Mn—0.50 to 0.70

Drop hammer Dies

Si—1.25 to 1.50
S—max 0.07
P—0.20
Mn—0.60 to 0.80

Locks & Hinges

Si—2.50 to 2.70
S—max 0.08
P—0.70 to 1
Mn—0.40 to 0.60

Soil Pipe & Fittings

Si—1.75 to 2.25
S—max 0.10
P—0.50 to 0.80
Mn—0.50 to 0.80

Steam Radiators

Si—2 to 2.25
S—max .08
P—0.40 to 0.60
Mn—0.50 to 0.70

Novelty Work(toys etc.)

Si—2.50 to 3
S—max .08
P—0.8 to 1
Mn—0.40 to 0.60

Fluxes, Pickles and Dips

The following is a collection of fluxes, pickles, dips and furnace linings that I have collected over the years. All have been foundry tested.

Bright Dip for Aluminum Castings

Dip in diluted solution of caustic soda, then dip into a solution of 50% nitric acid and 50% water. Wash in clear water and dry in hardwood sawdust.

Pickles for Brasses:

- Nitric Acid...1½ parts
  Sulphuric Acid...2 parts
  Sodium Chloride..2 ounces for each
     4 gal.
- Sulphuric acid ....5 parts
  Saltpeter..1 part
  Water...2.5 parts
- Yellow nitric acid...1 part
  Sulphuric acid...1 part
  Sodium chloride ....½ ounce per gal.
  Rinse in clean hot water containing a small amount of cream of tartar, dry in hardwood sawdust.
- Plastic Refractory:
  Brick grog...70 percent
  Fire Clay ....30 percent
  Mix dry before adding
     water.
- Brass Furnace Lining:
  Wisconsin ganester ⅜ inch mesh.................................50 percent
  Wisconsin ganster ⅛ inch mesh ....30 percent
  Silica flour ....14 percent
  Bentonite ....6 percent
- Castable Refractory:
  Grog #3 mesh ....70 percent
  Aluminite cement ....25 percent
  Fire clay...5 percent
- Core Wash:
  11 parts iron oxide, talc, or plumbago
  1 part bentonite
  Molasses water mixed to correct Baumé

Mold Spray:

Glutrin water to which graphite or talc is added to correct Baumé. Good for ferrous and nonferrous.

Brass Fluxes:

- 100 pound dehydrated borax
  77 pound whitting
  50 pound sodium sulphate
- 50 percent Razorite
  50 percent Soda ash
- 8 parts flint glass
  1 part calcined borax
  2 parts fine charcoal
- 5 parts salt
  5 parts sea coal
  15 parts sharp sand
  20 parts bone ash
- 25 percent Soda ash
  25 percent plaster of Paris
  25 percent fine charcoal
  25 percent salt
- 50 pound glass
  55 pound Razorite
  5 pound lime

Core Paste

50 percent foundry molasses
50 percent talc.

Old Timer Flux for Copper:

Mix equal parts of charcoal and zinc with enough molasses water to form a stiff paste. Roll into balls about 2 inches in diameter, then dry. When the copper just starts to melt drop in enough balls to give a good cover.

Tough-Job Core Wash:

4 parts silica flour
1 part delta core wash (50° Baumé for dipping & 40° for spraying)

Brazing Flux:

50 percent boric acid
50 percent sodium carbonate.

German Silver Flux:

5 parts ground marble
3 parts sharp sand
1 part borax
1 part salt
10 parts fine charcoal

Bright Dip for Copper Castings:

10 gallons water
1 gallon sulphuric acid
10 pound potassium dichromate

    Wash after bright dip and neutralize acid by dipping in a 1 percent sodium carbonate solution, then a final rinse.

General Mixes

Aluminum Bronze	**Brazing Metal**	**P.S. (Plumber's Special)**
Cu—90 percent	Cu—84-86 percent	Cu—80 percent
Al—10 percent	Zn—14-16 percent	Pb—6 percent
	Fe—0.06 percent maximum	Sn—3 percent
	Pb—0.30 percent	Zn—11 percent
		semired bronze

Grille Metal	**Journal Bronze**	**Imitation Manganese Bronze**
Cu—69 percent	Cu—82-84 percent	Cu—59 pounds
Sn—1 percent	Sn—12.5-14.5 percent	Zn—40 pounds
Zn—30 percent	Zn—2.5-4.5 percent	Al—5 ounces
	Fe—0.06 max	Sn—5 ounces
	Pb—1 max	Tin plate clippings—6 ounces

Cheap Mix for Plumber's Ware

Sheet metal clippings—71½ percent
Copper wire—25 percent
Pb—2 percent
Sn—1 percent
P Cu (15 percent) 8 ounces per
100 lb of above

Cheap Red Brass

Scrap copper wire—40 pound
Zinc—7.5 pounds
Lead—7.5 pounds
Mixed brass scrap—45 pounds
Melted mixed brass scrap add
copper, zinc and lead

Another Red Brass for Small Castings

Cu—82 percent	Zn—8 percent
Pb—8 percent	Sn—2 percent

Not used for steam work

Muntz (General Spec)

Cu—59-62 percent
Zn—39-41 percent
Pb—0.6 maximum

Hard Bearing Bronze	**Commercial Yellow Bronze**	**Statuary Mix #1**
Cu—63.5 percent	Cu—69 percent	Cu—90 percent
Zn—21.5 percent	Pb—3 percent	Sn—7 percent

Hard Bearing Bronze

Manganese—4 percent
Fe—4 percent
Al—7percent

Commercial Yellow Bronze

Sn—1.5 percent
Zn—26.5 percent

Statuary Mix #1

Zn—3 percent
8 ounces of Pb per
100 pounds

Statuary Mix #2

Cu—90 percent
Sn—5 percent
Zn—5 percent
8 ounces of Pb per
100 pounds

Statuary Mix #3

Cu—90 percent
Zn—7.5 percent
Sn—2.5 percent
Also used as a tablet alloy

Statuary Mix #4

Cu—88 percent
Sn—6 percent
Zn—3.5 percent
Pb—2.5 percent

Commercial Brass

Cu—64-68 percent
Zn—32-34 percent
Fe—2 percent max
Pb—3 percent max

Gun Bronze

Cu—87-89 percent
Sn—9-11 percent
Zn—1-3 percent
Fe—0.06 percent maximum
Pb—0.30 percent maximum

Ornamental Bronze

Cu—83 percent
Pb—4 percent
Sn—2 percent
Zn—11 percent

Red Ingot

Cu—85 percent
Pb—5 percent
Sn—5 percent
Zn—5 percent

Pressure Metal

Cu—83 percent
Pb—7 percent
Sn—7 percent
Zn—3 percent

Gear Bronze

Cu—87.5 percent
Pb—1.5 percent
Sn—9.5 percent
Zn—1.5 percent

Bronze

Cu—89.75 percent
Sn—10 percent
Ph—0.25 percent

88-4 Bronze Ingot

Cu—88 percent
Sn—8 percent
Zn—4 percent

Bearing Bronze

Cu—80 percent
Pb—10 percent
Zn—10 percent

Lead Lube

Cu—70 percent
Pb—25 percent
Sn—5 percent

Manganese Bronze

Cu—60 percent
Zn—42 percent
Balance temper depending
on grade desired

Nickel Silver

Cu—61 percent
Zn—20 percent
Ni—18 percent
Fe—1 percent

Ship Bells

Cu—82 percent
Sn—12 percent
Zn—6 percent

Cheap Yellow Brass

Cu— 60 percent
Zn—36 percent
Pb—4 percent

Another Cheap Red Brass

Cu—83 percent
Pb—8.5 percent
Zn—6.5 percent
Sn—2 percent

Yellow Brass for Small Castings

Cu—70 percent Sn—2.8 percent
Zn—25.5 percent Pb—2 percent

Half & Half Yellow-Red Brass

Cu—55 percent Zn—10 percent
Pb—5 percent Yellow brass
scrap 30 percent

Tough Free Bending Metal

Cu—84.5 percent
Zn—10 percent
Pb—3 percent
Sn—2.5 percent

Bell Mixes

(White) Table Bells

Sn—97 percent
Cu—2.5 percent
Bi—0.5 percent

Swiss Clock Bells

Cu—75 percent
Sn—25 percent

Silver Bells

Cu—40 percent
Sn—60 percent

Best Tone

Cu—78 percent
Sn—22 percent

House Bells

Cu—78 percent
Sn—20 percent
Yellow Brass—2 percent

Sleigh Bells

Cu—40 percent
Sn—60 percent

High Grade Table Bell

Sn—19 percent
Ni—80 percent
Pt—1 percent

Special Clock Bells

Cu—80 percent
Sn—20 percent

Special Silver Bells

Cu—50 percent
Zn—25 percent
Ni—25 percent

General Bell

Cu—80 percent
Sn—20 percent

Fire Engine Bells

Sn—20 percent
Ni—2 percent
Balance Cu

Large Bells

Cu—76 percent
Sn—24 percent

Railroad Signal Bells

Cu—60 percent
Zn—36 percent
Fe—4 percent

House Bells

Cu—76 percent
Sn—16 percent
Yellow Brass—8 per-cent

Gongs

Cu—82 percent
Sn—18 percent

Miscellaneous Metal Mixes

Slush Metal

Zinc..............92%
Aluminum........5%
Copper 3%

Oriental Bronze

Copper 84%
Lead 10%
Tin....................5%
Zinc...................1%

Imitation Gold

Copper..............89%
Zinc 10.5%
Aluminum..........0.5%

Aluminum Bronze

Copper 90%
Iron 05%
Aluminum................9%
Manganese Copper...05%

Phosphor Bronze

#1 copper wire 90%
Phosphor Tin 10%

Imitation Silver

Copper...............50%
Nickel 20%
Zinc...................30%

High Pressure Bronze

Copper 84%
Tin 7.5%
Lead 5.5%
Zinc.......................3%

Casket Metal Trim

Tin....................0.5%
Antimony............12.5%
Arsenic................8.5%
Lead 87.5%

Aluminum Solder

Tin....................68%
Antimony 0.5%
Zinc 21.5%

408

Approximate Analysis Various Grades Scrap Aluminum

Material	Cu	Fe	Zn	Si	Misc
Crankcase	7-8	1-1.25	1-1.5	1-1.5	
Misc Cast	7-8	1-1.25	1-1.5	1-2	
Cable	0-0.10	.2-.50	0-0.2	0.2-0.4	
Pistons	8-9	0.76-1.25	0.1-0.5	0.75-1.25	
New Clips	0.1-0.3	0.3-0.8	0-0.2	0.2-0.7	
Alum Die Cast	6-7	1-1.5	1-2	2-3	Ni 0.5-1.0
Dural Clips	4-4.5	0.5-1.	0-0.25	0.25-0.5	Mg 0.5-1.0
					Mn 0.5-1.5
Pots-Pans Dishes	0.25-0.50	0.50-0.60	0.25-0.50	0.25-0.50	

Judging the Approximate Temperature of an Object in the Dark by Eye:

Faint Red ....878°F
Dull Blood Red ....990°F
Full Blood Red ....1050°F
Dull Cherry Red ....1195°F
Full Cherry ....1375°F
Light Cherry..1550°F
Deep Orange ....1640°F
Light Orange ....1730°F
Yellow...1832°F
Light Yellow ....1976°F
White...2200°F
Bright White..2550°F
Dazzling White ....2730°F

Investments and Their Application

Commercially Prepared Investments	Mold Construction	Casting Metals	Max. Pouring Temp.
Hydro Perm (U. S. Gypsum Co.)	Flask Mold Full Mold	Brass Bronze Aluminum	2000°F.
Investment R Investment R 555 (Ransom & Randolph Co.)	Flask Molds Hand-build molds	All (below max. pouring temp)	2000°F.
Duracast 20 (Kerr Mfg. Co.)	Flask Molds Hand-built Molds	All (below max. pouring temp.)	2000°F.
No. 1 Molding Plaster (Georgia Pacific Inc.)	All	All (below max. pouring temp).	2000°F.
Noncommercial Investments 95% talc		Brass	2000°F.
4% Hi-Early cement 1% asbestos (by weight)	Full molds	Aluminum Bronze	
Silica sand and fire clay (1:1 by volume)	All	all	2000°F.
70% No. 1 Molding Plaster 29% talc 1% hydrated lime 0.3% portland cement (by weight).	All	Brass Bronze Aluminum	2000°F.
54% 100-mesh silica sand 32% No. 1 molding plaster 13% talc 1% Hi-Early cement (by weight).	All	Brass Bronze Aluminum	2000°F.
200-mesh olivine flour and No. 1 molding plaster (3:2 by volume)	All	Brass Bronze Aluminum	2000°F.
Calcined plaster of paris (gypsum)	Flask molds Full molds	Aluminum lead	(none)
Plaster of paris and silica flour (1:1 by volume)	All	Brass Bronze Aluminum	2000°F.
Brick dust and plaster of paris (3:2 by volume)	All	Brass Bronze Aluminum	2000°F.
100-mesh silica sand and Hi-Early cement (10:1 by volume)	Flask molds Full molds	All	2000°F.

Average Analysis of Alloy Scrap (%)

	Cu.	Sn	Pb	Zn	Fe	Ni	Sb	Slag	Si
No. 1 copper	99.00(min.)	-	-	-	-	-	-	-	-
No. 2 copper	95.00(min.)	0.5	0.5	-	-	-	-	-	-
Heavy copper	99.00	0.5	0.5	-	-	-	-	-	-
Light copper	92.09	1.00	2.00	-	-	-	-	-	-
Oillite bushings	90.0	10.0	-	-	-	-	-	-	-
No.1 high grade bronze castings	88.0	7.5	0.5	bal	-	-	-	-	-
No.2 high grade bronze castings	88.0	6.0	2.0	bal	-	-	-	-	-
Phosphor—bronze	95.0	5.0	-	-	-	-	-	-	-
Bell metal	81.0	17.6	0.6	bal	-	-	-	-	-
	Cu.	Sn	Pb	Zn	Fe	Ni	Sb	Slag	Si
Red scrap & turnings	85.0	4.5	5.0	bal	-	-	-	-	-
Yellow scrap turning	70.0	1.0	2.0	bal	0.6	-	-	-	-
Ni-plated brass tubing	67.0	0.1	0.1	bal	-	-	0.6	-	-
Boiler condenser tubes ⅞in aluminum	77.7	-	-	bal	-	-		2.5	-
Admiralty metal	70.0	0.9	-	bal	-	-	-	-	-
Muntz metal	63.0	-	0.2	bal	-	-	-	-	-
Yellow brass faucets	70.00	-	-	bal	-	-	-	-	-
Yellow rod stock	62.00	-	2.00	ba;	-	-	-	-	-
Light brass	67.00	1.00	1.00	nal	0.50	-	-	-	-
Rolled brass	67.00	0.1	0.1	bal	-	-	-	-	-
Auto oil coolers, round	80.50	3.95	4.00	bal	-	-	-	-	-
Automobile radiators with copper tanks	73.50	4.85	17.75	bal	-	-	-	-	-
	73.50	4.85							
Automobile radiators with brass tanks	66.09	6.75	15.25	bal	0.40	-	-	-	-
Railroad wheel bearings	75.7	5.50	17.00	bal.	-	-	0.60	-	-
Ni-Cu Scrap	70.00	-	-	-	-	30.00	-	-	-
Monel metal clippings	34.0	-	-	-	1.00	65.00	-	-	-
Ni-Ag scrap	bal.	1.00	1.00	15.00	1.00	18.50	-	-	-
Mixed Si-bronze	95.00	-	-	2.00	1.00	-	-	-	2.00
Al-bronze	88.0	-	-	-	3.0	-	9.0		
Duronz	92.0	-	-	-	1.0	-		3.0	2.00
85/15 pipe	85.8	-	-	bal.	-	-	-	-	-
No. 1 Babbit or pewter		80.0	10.0	-	-	-	10.0	-	-
Manganese-bronze	60.0	0.5	0.5	bal.,	1.0	-	-	1.0	

Foundry Suppliers

FOUNDRY FACINGS & SUPPLIES

**C. E. Cast Industrial Products
Combustion Eng. Inc.**
14365 Wyoming Ave.
Detroit, Mich. 48238

Refractories and refractory materials

C. M. Smillie & Co.
1180 Woodward Heights Blvd.
Ferndale, Michigan 48220

Foundry and plastic supplies

Certech Inc.
53-57 Bergentine Ave.
Westwood, N. J. 07675

Injection molded ceramics

Ferro Corp., Electro Div.
661 Willet Road
Buffalo, N. Y. 14218

Refractory strainer cores, riser tile, runner, gate tile and breaker cores

Akron Porcelain Co.
P. O. Box 3767
Akron, Ohio

Foundry strainer cores

Universal Clay Products Co.
1500 First St.
Sandusky, Ohio 44870

Refractory gating components

**Diamonite Products Div.
of Spartex**
P. O. Box TR
Shreve, Ohio 44676

Aluminum oxide ceramic products, sand blast nozzles

**C. E. Cast Industrial
Products Arcoa Div.**
4401 Creekside Ave.
Toledo, Ohio 43612

Refractory core, mold coatings and binders

Du-Co Ceramics Co.
Mill Run Rd.
Saxonburg, Pa. 16056

Ceramic insulation, high temp. refractories, etc.

Saxonburg Ceramics Inc.
100 Isabelle St.
Saxonburg, Pa. 16056

Refractories, strainer cores crucibles, etc.

FOUNDRY FACINGS & SUPPLIES

T. H. Benners & Co.
701 S. 37th
Birmingham, Ala. 35222

Foundry raw materials, etc.

Huntington Beach Plant, Electro Div. Ferro Corp.
P. O. Box 471
Huntington Beach, Calif. 92648

Refractories, crucibles, restrainer cores etc.

Beaver Cut Rotary File Co.
Telegraph Rd. & Woods
Los Angeles, Cal. 90022

Foundry supplies, files, rasps, etc.

Electro-Coatings Inc.
1601-05 School St.
Moraga, Cal. 94556

Mold coats

Perfect Plank Co.
1559 Orobrodam Blvd.
Oroville, Cal. 95965

Wood Laminate for Patterns

Asbury Graphite Inc. of Cal.
2855 Franklin Canyon Rd.
Rodeo, Cal. 94572

All types of graphite

Casting Materials Co. Inc.
4614 S. Hampton St.
Vernon, Cal. 90058

Foundry supplies

Capewell Div. Stanadyne Inc.
61 Governor St.
Hartford, Conn. 06102

Saw blades, chill nails

John R. Neilson & Sons Inc.
252 Chapel Rd.
South Windsor, Conn. 06074

Wood flour of all types

Composition Material Inc.
26 Sixth St. Dept. TR
Stamford, Conn. 06905

Wood flour, fibers, etc.

Equipment Co. of America
1077 Hialeah Drive
Hialeah, Fla. 33010

Magnesium bottom boards

Walker Peenimpac Div. of Walker Pump Co. *Shot and Grit*
2010 E. Hill Ave. Box 755
Valdosta, Ga. 31601

Atlantic Chemical & Metals Co. *Chemical fluxes*
1925 N. Kenmore Ave.
Chicago, Ill. 60614

Barco Chemical Products Co. *Aerosol silicone mold release*
703 S. La Salle St.
Chicago, Ill. 60605

Black Products Co. *Foundry core compounds, facings*
13515 S. Calumet Ave.
Chicago, Ill. 60627

J. R. Short Milling Co. *Foundry equipment*
233 S. Wacker Dr.
Chicago, Ill. 60606

Superior Graphite Co. *Graphite*
20 N. Wacker Dr.
Chicago, Ill. 60606

Moulder's Friend Inc. *Foundry sand*
Thomas St. *conditioners and aerators*
Dallas City, Ill. 62330

Midwest Foundry Supply Co.
Edwardsville, Ill. 62025

Acme Resin Co. *Resins and binders*
14 01 Circle Ave.
Forest Park, Ill. 60130

Smith & Richardson Mfg. Co. *Foundry chaplets and supplies*
705 May St.
Geneva, Ill. 60134

Gallagher Corp. *Foundry plastics*
3966 Morrison Ave.
Gurnee, Ill. 60031

Ipsen Ceramics *High temperature*
330 John St. *crucibles, etc.*
Pecatonica, Ill. 61063

Coeval Inc. *Corn cob flour*
St. Joseph, Ill. 61873

American Colloid Co. *Pure Wyoming*
5100 Suffield Court, Dept. TB *bentonite*
Skokie, Ill. 60076

Solomon Grinding Service
P. O. Box 1766
Springfield, Ill. 62705

Iron oxide

Pitner & Hess Inc.
Bedford, Indiana 47421

*Organic non-mineral
foundry parting*

Indiana Products Co.
La Salle West Bldg.
South Bend, Ind. 46601

*Foundry facing
and supplies*

**Carver Foundry
Products Inc.**
Capital Road, Progress Park
P. O. Box 817
Muscatine, Iowa 52761

Supplies and machinery

M. W. Hartman Mfg. Co.
400 W. Second
Hutchinson, Kansas 67501

*Foundry machinery
and equipment*

Foundry Rubber Inc. Para Products Div
7841 Airpark Rd.
Gaithersburg, Md. 20760

Core binders, liquid partings, etc.

Eastern Wood Fibers Inc.
8245 Dorsey Run Rd.
Jessup, Md. 20794

Wood flour etc.

Devcon Corp.
59 Endicott St.
Danvers, Mass. 01923

Plastic steel

Boston Pattern Works Inc.
595 Pleasant
Norwood, Mass. 02082

Castings and pattern equipment

**Bullard Abrasive
Products Inc.**
40 Donald St.
Westboro, Mass. 01581

Grinding and cut-off wheels

Foundries Materials Co.
5 Preston St.
Coldwater, Michigan 49036

Foundry supplies

**Colloidal Paint
Products Co.**
6136 Charles St.
Detroit, Michigan 48212

*Graphoidal coatings
for metal mold*

**Durez-Stevens
Foundry Supply**
810 Rosewood
Ferndale, Mich. 48220

*Core wash, paste,
plumbago, parting etc.*

Spurgeon Co.
1330 Hilton Rd.
Ferndale, Mich. 48220

Conveyors, etc.

Chem Trend Inc. Dept. T.
3205 E. Grand River
Howell, Michigan 48843

Mold releases

Roto-Finish Co.
3707 Milham Road
Kalamazoo, Mich, 49003

Finishing and deburring machines, abrasive media

Acheson Colloids Co.
1637 Washington Ave.
Port Huron, Mich. 48060

*Refractory and
colloidal graphite*

**Severance Tool
Industries Inc.**
3790 Orange St.
P. O. Box 1866-P
Saginaw, Mich. 48605

*Ground high speed
and carbide midget
mills for removing fins, etc.*

**Wickes Engineered
Materials Div.**
1621 Holland Ave.
Saginaw, Mich. 48601

*Sand conditioner
and ingot mold wash*

**Grav-I-Flo Corp.
Norwood and Davidson**
Sturgis, Mich, 49091

Deburring equipment

Mecha-Finish Corp.
Box 308-T
Sturgis, Mich. 49091

*Automatic de-
burring equipment*

Acme Abrasive Co.
24202 Marmon Ave.
Warren, Michigan 48089

Grinding wheels, etc.

Simpson Russ Co.
21908 Schoenherr Road
Warren, Mich. 48089

Pattern supplies

416

Sterling Machine & Supply Co.
2222 Elm S. E.
Minneapolis, Minn. 55414

Foundry supplies

Technical Ceramics Products Div.
3M Center
St. Paul, Minn. 55101

Strainer cores

Products Engineering Co.
Cape Girardeau, Mo. 63701

Foundry jackets and liners

Asbury Graphite Mills Inc.
41 Main St.
Asbury, N. J. 08802

Foundry facings, blackings and plumbago

Whitehead Brothers Co.
66 Hanover Rd.
Florham Park, N. J. 07932

Foundry sand and supplies

Springfield Facing Corp.
401 S. 2nd St.
Harrison, N. J. 07029

Foundry facings and supplies

Lignum Chemical Works
No. 5-foot of Jersey Ave.
Jersey City, N. J.

Sawdust, wood flour

**Belmont Metals Inc.
Div. of Belmont Metal & Refining Wks.**
320 Belmont Ave.
Brooklyn, N. Y. 11207

Smelters and refiners of all nonferrous metals

Sirotta Bernard Co., Inc.
65—35th St.
Brooklyn, N. Y. 11232

Nut shell blast cleaning abrasives, etc.

United Compound Co.
611 Indian Church Rd.
Buffalo, N. Y. 14224

Vent wax

Remet Corp.
Dept. A. P. O. Box 278
278 Bleachery Place
Chadwicks, N. Y. 13319

Foundry facings and supplies

**Great Lakes Carbon
Corp. Graphite Products Div.**
299 Park Ave.
New York, N. Y. 10017

*Graphite powders,
chill rods fluxing tubes, etc.*

**Liberty Steel
& Metals Corp.**
Time & Life Bldg.
Rockefeller Center
New York, N. Y. 10020

*Copper shot, master
alloys, ferro alloys*

Pancoast International Corp.
3-5 Park Row
New York, N. Y. 10038

*Copper shot, master
alloys, ferro alloys*

**NL Industries Inc.,
Tam Div.**
Box c., Bridge Station
Niagara Falls, N. Y. 14305

*Zirconium oxides,
silicates and chemicals*

Unicast Development Corp.
345 Tompkins Ave.
Pleasantville, N. Y. 10570

*Cer. and high temperature
facings core and mold binders*

**Cleveland Metal
Abrasive Inc.**
305 Euclid Office Plaza
Cleveland, Ohio 44101

Blast cleaning abrasives

The Exolon Company
950 E. Niagara St.
Tonawanda, N. Y. 14150

*Aluminum oxide and
silicon carbide abrasive, etc.*

Polymer Applications Inc.
3443 River Rd.
Tonawanda, N. Y. 14150

*Shell mold and
cold set resins*

Precise Alloys Inc.
69-T Kinkel St.
Westbury, N. Y. 11590

*Solders, lead,
lead castings etc.*

Nordson Corp.
314 Jackson St.
Amherst, Ohio 44001

Hot paste applicators

Buckeye Products Co.
7021 Vine St.
Cincinnati, Ohio 45216

*Foundry supplies
and equipment*

Hill & Griffith Co.
1265-1267 State Ave.
Cincinnati, Ohio 45204

*Foundry facings
and supplies*

BMM Inc.
7777 Wall St.
Cleveland, Ohio 44125

Foundry equipment

Cleveland Metal Abrasive Inc.
305 Euclid Office Plaza
Cleveland, Ohio 44101

Abrasives for blast cleaning castings, etc.

Foseco Inc.
P. O. Box 81227
Cleveland, Ohio 44181

Foundry coatings and flux, etc.

Inland Refractories Co.
10235 Berea Rd.
Cleveland, Ohio 44102

Cupola gun mix, brick, clay etc.

Kindt-Collins Co.
12655 Elmwood Ave.
Cleveland, Ohio 44111

Rapping plates, pattern supplies, wax products, etc.

Metal Blast Inc.
871 E. 67th St.
Cleveland, Ohio 44103

Iron and steel shot and grit

**Smith Facing
& Supply Co.**
1859 Carter Road
Cleveland, Ohio 44113

*Grinders of
seacoal, etc.*

W. S. Tyler Inc.
C. E. Equipment Div.
7887-T Hub Pkwy.
Cleveland, Ohio 44125

*Mold washes, core binders sand, mold
coatings, etc.*

Ashland Chemical Co.
P. O. Box 2219
Columbus, Ohio 43216

Binders, seacoal, core and mold washes, etc.

Borden Chemical Div. of Borden Inc.
180-T E. Broad St.
Columbus, Ohio 43215

Binders, adhesives, resins, etc.

Exomet Inc.
Box 647-T
Conneaut, Ohio 44030

*Insulating products,
core and mold castings*

Louthan Plant
Div. Ferro Corp.
P. O. Box 781
East Liverpool, Ohio 43920

Gating and risering tile,
strainer cores, refractory
specialties, etc.

Globe Steel Abrasive Co.
Laird & First Ave.
Mansfield, Ohio 44901

Steel and iron shot
and grit for blast cleaning

The Andersons Cob Division
519 Illinois Ave.
Maumee, Ohio 43537

Corn cob flour

Brown Insulating
Systems Inc.
11943 Abbey Rd.
North Royalton, Ohio 44133

Insulating riser shapes, etc.

Rudow Mfg. Co.
Sandusky, Ohio 44870

Shell sand runner
tubes and pouring basins

Universal Clay Products Co.
1500 First St.
Sandusky, Ohio 44870

Refractory gating components

Barium & Chemicals Inc.
Steubenville, Ohio 43952

Foundry chemicals

Freeman Mfg. Co.
1150 E. Broadway
Toledo, Ohio 43605

Pattern shop and foundry supplies and equipment

Ransom & Randolph Co.
Chestnut & Superior Sts.
Toledo, Ohio 43604

Investments

George F. Pettinos Inc.
123 Coulter Ave.
Ardmore, Pa. 19003

Sands and gravel
refractory cements,
graphites, etc.

C.E. Cast Industrial Products
4401 Creekside Ave.
Toledo, Ohio 43612

Permanent mold coatings.
partings, sealants, etc.

McGee Industries Inc.
9 Crozerville Rd.
Aston, Pa. 19014

Fluorocarbon mold release compound

Limewood Corp.
R. D. 1
Boyers, Pa. 16020

Foundry facings

United Erie Inc.
1429 Walnut St.
Erie, Pa. 16512

Resins, core oils, grease, etc.

Pennsylvania Foundry Supply And Sand Company
6801 State Road
Bldg. B.
Philadelphia, Pa. 19135

Abrasive Metals Co.
2602 Smallman St.
Pittsburgh, Pa. 15222

*Chilled and Malleable
shot and grit, etc.*

Durasteel Abrasive Co.
2602 Smallman St.
Pittsburgh, Pa. 15222

Electric arc furnace cast steel shot and grit

**Foundry Warehousing
& Supply Co.**
A.V.R.R. & 33rd St.
Pittsburgh, Pa. 15201

Facings and supplies

Fuel Equipment Co.
104 Fourth St.
Pittsburgh, Pa. 15215

*Heaters, dryers,
gas burners*

**Pittsburgh Metals
& Purifying Div.**
P. O. Box 260
Saxonburg, Pa. 16056

*Exothermic and
insulating compounds*

Shamokin Filler Co.
Shamokin, Pa. 17872

*Carbon fillers
and additives*

Grafos Colloids Corp.
299 Wilkes Place
Sharon, Pa. 16146

*Manufacturers colloidai
graphite lubricants, etc.*

M. E. Wallace Co.
Carbon St.
Sunbury, Pa. 17801

Carbon Compounds

Watsontown Products Co.
Watsontown, Pa. 17777

*Fillers, foundry
facings, etc.*

Allegheny Enterprises
Industrial Park
Zelienople, Pa. 16063

Metallurgical chemicals, foundry supplies

Porter Warner Industries Inc.
3819 Dorris St.
Chattanooga, Tenn. 37410

Molding sands, glass sands, clays, talcs

ACF Plant Electro Div. Ferro Corp.
E. Duncan St.
Tyler, Texas 75701

Gating and risering tile, strainer cores

Vermont Talc Inc.
46 Mill St.
Chester, Vermont 05143

Talc and soapstone facings

Blue Ridge Talc Inc.
Henry, Va. 24102

Talc and soapstone facings

Northwest Olivine
5515 Fourth Ave. S.
Seattle, Washington 98134

Olivine

Technical Specialties International
487 Elliott Ave. W.
Seattle, Washington 98119

*Precision casting waxes,
injection mold waxes*

**Donald Sales
& Mfg. Co.**
6601 W. State
Milwaukee, Wisc. 53213

*Seacoal, pitch
core comp.*

Merit Corp.
2400 S. 43rd
Milwaukee, Wis. 53219

Foundry facings and supplies

Solvox Mfg. Co.
11727 W. Fairview Ave.
Milwaukee, Wis. 53226

Metal cleaners, foundry fluxes

Thiem Corp.
9800 W. Rogers
Milwaukee, Wis. 53227

Foundry facings and compounds

Milwaukee Chaplet & Mfg. Co. Inc.
17000 W. Rogers Drive
New Berlin, Wis. 53151

Chaplets and chills

Ace Mfg. Co.
Mary & Wisconsin Sts.
Weyauwega, Wis. 54983

Slotted brass and aluminum core box vents, etc.

FOUNDRY PLANT EQUIPMNET

**Branford Vibrator Co. Div. of
Electro Mechanics Inc.**
152 John Downey Dr.
New Britain, Conn. 06051

*Pneumatic and electric
operated foundry vibrators*

QC Inc.
1450 Southwest 12th Ave.
Pompano Beach, Fla. 33060

New and used blast cleaning equipment

Abrading Machinery & Supply Co.
2336 W. Wabsania Ave.
Chicago, Ill. 60647

Barrel and vibrating equipment

Whiting Corporation
15700 Lathrop Ave.
Harvey, Ill. 60426

*Complete foundry
equipment*

Chemaperm Magnetics Inc.
230-T Crossways Park Dr.
Woodbury, N.Y. 11797

Magnetic plates, pulleys and drums

Martin Engineering Co.
Rte. 34 Dept. TR
Neponset, Ill. 61345

Foundry Equipment

Kassnel Vibrator Co.
5990 N. River Rd.
Rosemont, Ill. 60018

Foundry vibrators

The Adams Co.
Philip & Adams Sts.
Dubuque, Iowa 52001

Flasks, jackets, bands, bottom plate, sprue cutters

Ohler Machinery Co
100 Industry Ave.
Waterloo, Iowa 50704

Small inexpensive core sand mixer

Tex Specialities Inc.
265 Swanton St.
Winchester, Mass. 01890

Coreless induction furnace, etc.

Keyes-Davis Co.
64 14th St.
Battle Creek, Mich. 49016

Aluminum bottom boards

Roberts Corp.
P.O. Box 13160T
Lansing, Mich. 48901

Foundry equipment

Crown Iron Works Fdry. Equip. Division
P.O. Box 1364
1300 Tyler St. NE
Minneapolis, Minn. 55440

Sand mixers, foundry equipment

Pyro-Serv Instruments
Locust Ave. & River Rd.
North Arlington, N.J. 07032

Hand pyrometers

Inductotherm Corp.
12 Indel Ave.
Rancocas, N.J. 08073

*Induction melting and
heating equipment*

Alexander Saunders & Co.
Route 301
Cold Spring, N.Y. 10516

*Investment cast-
ing equipment, foundry
equipment and supplies*

Haywood Supply Co. Inc.
540 S. Columbus Ave.
Mt. Vernon, N. Y. 07207

Sand handling

Casting Supply House Inc.
62 West 47th St.
New York, N.Y. 10036

*Nonferrous
investment, wax
foundry supplies*

Unicast Development Corp.
345 Tomphins Ave.
Pleasantville, N.Y. 10570

*Casting equipment, foundry,
facings and supplies*

The Macleod Co.
11125 Mosteller Rd.
Cincinnati, Ohio 45241

*Metal scrap washing and
cleaning machinery*

Young & Bertke Co.
2213-15 Winchell Ave.
Cincinnati, Ohio 45214

*Dust and fuel
control systems*

American Monorail
P.O. Box 4338T
Cleveland, Ohio 44143

*Hot metal delivery and
pour-off systems*

Cleveland Products Inc.
P.O. Drawer 2754T
Cleveland, Ohio 44111

*Green sand mold-
ing systems, molding,
coremaking and mold
handling equipment*

Marketeers Inc.
19101 Villa View Road
Cleveland, Ohio 44119

*Investment casting
equipment and supplies*

Rose Metal Industries Inc.
1538 E. 43rd St.
Cleveland, Ohio 44103

Skimmers,
tongs, ladles in stock

Frank A. Walsh & Sons Inc.
13004 Athens Ave.
Cleveland, Ohio 44107

Sand blast equipment,
tumbling mills, dust collecting systems

Jeffrey Mfg. Div. Dresser Ind.
P.O. Box 2251-T
400 W. Wilson Bridge Rd.
Columbus, Ohio 43216

Flask handling,
continuous mullers, mold conveyors

Whitehead Brothers Co.
66 Hanover Road
Florham Park, N.J. 07932

Foundry sand
and supplies

Beardsley & Piper Div. of Pettibone
5503 W. Grand Ave.
Chicago, Ill. 60639

Sandslingers, sand conditioning and
molding equipment

FOUNDRY METALS

American Smelting & Refining Co.
120 Broadway
New York, N.Y. 10005

Brass, bronze, aluminum ingots,
deoxidizers, phos. copper, lead, copper, etc.
Sales offices in 15 major U.S. cities and Canada.

Belmont Smelting
& Refining Works Inc.
330 Belmont Ave.
Brooklyn, N.Y. 11207

Brass, bronze and aluminum ingots,
deoxiders, hardeners, white metals, etc.

H. Kramer & Co.
1315 W. 21st St.
Chicago, Ill. 60608

Brass, bronze and
aluminum ingots, etc.

I. Schumann & Co.
22501 Alexander Rd.
Cleveland, Ohio 44146

Brass and bronze ingots

Sitkin Smelting & Refining Co.
P.O. Box 708
Lewistown, Pa. 17044

Brass, bronze ingots, etc.

C.F. & I. Steel Corp.
P.O. Box 1830
Pueblo, Colo. 81002

Foundry pig iron

Interlake Inc.
135th St. & Perry Ave.
Chicago, Ill. 60627

Foundry pig iron

425

Jackson Iron & Steel Co.   *Foundry pig iron*
Lick Township
Jackson, Ohio 45640

Lone Star Steel Co.   *Foundry pig iron*
2200 W. Mockingbird Lane
Dallas, Texas 75235

National Steel Corp.   *Foundry pig iron*
2800 Grant Bldg.
Pittsburgh, Pa. 15219

Reynolds Metals Co.   *Aluminum ingots*
P.O. Box 27003ZA
Richmond, Va. 23261

Alcoa   *Aluminum ingots*
1222 Alcoa Bldg.
Pittsburgh, Pa. 15219

Kayser Aluminum   *Aluminum ingots*
6250 E. Banbini Blvd.
Los Angeles, Calif.

MOLDING MACHINES

Bradford Vibrator Co.   *Div. Electro*
150 John Downey Dr.   *Mechanics*
New Britain, Conn. 06051

The Osborn Mfg. Co.
5401 Hamilton Ave.
Cleveland, Ohio 44114

Mac Erie Mfg. Co.
1114 Walnut St.
Erie, Pa.

Cleveland Products Inc.
P.O. Box 2754T
Cleveland, Ohio 44111

Tabor Industries Inc.
840 W. Main St.
Lansdale, Pa. 19446

National Air Vibrator Co.
6880 Wynnwood Lane
Houston, Texas, 77008

FOUNDRY SANDS

Arkhola Sand & Gravel
P.O. Box 1627
Ft. Smith, Ark. 72901

Del Monte Properties Co.
P.O. Box 567
Pebble Beach, Calif. 93953

Edgar Plastic Kaolin Co.
Box 8
Edgar, Florida 32049

Bell Rose Silica Co.
Central Life Bldg. at Ruby
Ottawa, Ill. 61350

Arrowhead Silica Corp.
Box 67
Chestertown, Ind. 46304

Hardy Sand Co.
Box 629
Evansville, Ind. 47701

Foundries Material Co.
5 Preston St.
Coldwater, Mich. 49036

Sargent Sand Co.
2844 Bay Rd.
Saginaw, Mich. 48603

Great Lakes Minerals Co.
2855 Coolidge Hwy. Suite 202-T
Troy, Mich. 48084

Inland Refractories Co.
10235 Berea Rd.
Cleveland, Ohio 44102

Penn Foundry Supply & Sand Co.
6801 State Rd. Bldg. B.
Philadelphia, Pa. 19135

American Gilsonite Co.
1150 Kennecott Bldg.
Salt Lake City, Utah 84133

MISCELLANEOUS FOUNDRY SUPPLIES & EQUIPMENT

Fox Grinders Inc.   *Grinders*
Harmony, Pa. 16037

Electromelt Corp.   *Furnaces*
32nd St. P.O. Box 4023
Pittsburgh, Pa. 15201

Vesuvious Crucible Co.
3636 Blvd. of the Allies
Pittsburgh, Pa. 15213

Cleveland Crane
Wickliffe, Ohio 44092

Clearfield Machine Co.   *Rotary furnaces*
Clearfield, Pa. 16830

International Molding Machine Co.
1201 N. Barnsdale Road
Lagrange Park, Ill. 60525

BMM Western Inc.   *Molding machines*
333 W. 78th
Minneapolis, Minn. 55420

Freemont Flask Co.   *Snap flasks*
1000 Wolfe Ave.
Freemont, Ohio 43420

Midmark Corp.   *Ladles*
Minster, Ohio 45865

Products Engineering Co.   *Miscellaneous*
Cape Girardeau, Mo. 63701

Stahl Speciality Co.   *Miscellaneous*
Kingsville, Missouri 64061

Cedar Heights Clay Co.   *Clay*
50 Portsmouth Rod.
Oak Hill, Ohio 45656

North American Refractories
900 Hanna Bldg.
East 14th & Euclid
Cleveland, Ohio 44115

3M Co.   *Abrasives, pattern supplies*
3M Center
St. Paul, Minn. 55101

Dupont Co.   *Foundry sands and chemicals*
Room 36357
Wilmington, Del. 19898

Electro-nite   *Miscellaneous*
Caroline Rd.
Philadelphia, Pa. 19154

American Induction Heating Corp.   *Ind.*
5353 Concord Ave.   *furnaces*
Detroit, Mich. 48211

Hanna Furnace Corp.
1299 Union Rd.
Buffalo, N.Y. 14224

Harry W. Dietert Co.
9330 Rose Lawn
Detroit, Mich. 48204

Sand testing equipment

Internation Minerals & Chemicals
Foundry products Div.
666 Garland Place
Des Plaines, Ill. 60016

Sand olivine

Ottawa Silica Co.
Box 577A
Ottawa, Ill. 61350

Sand

M.A. Bell Co.
217 Lomard
St. Louis, Mo. 63102

Miscellaneous

Jas. A Murphy & Co. Inc.
1421 E. High St.
Hamilton, Ohio 45011

Pistol sprayer

Sioux Tools Inc.
2901 Floyd Blvd.
Sioux City, Iowa 51102

Grinders etc.

GCA Corp.
Vacuum Ind. Division
34 Linden St.
Somerville, Mass. 02143

American Flask Co.
2745 South West Blvd.
Kansas City, Missouri 64108

Flasks

Hunter Automated Mach. Co.
2222 Hammond Drive
Schamburg, Ill. 60196

Molding mach. automatic

Brown Insulating Systems
11941 Abbey Rd.
Cleveland, Ohio 44133

Riser sleeves

Osborn Mfg. Corp.
5401 Hamilton Ave.
Cleveland, Ohio 44114

Molding machines

Brown Dover Corp.
North Brunswick, N.J. 08902

Induction melters

Union Carbide Corp. Metals Div.
270 Park Ave.
New York, N.Y. 10017

ferro alloys

Ajax Magnethermic Corp.
Warren, Ohio 44482

Induction melters

429

Modern Equipment Co. *Cupolas*
Box 266
Port Washington, Wisconsin 53074

Reichhold Chemicals Inc. Foundry Products Group *Foundry Resins*
525 North Broadway
White Plains, N.Y. 10603

Shalco Systems *Shell core machine*
12819 Coit Rd.
Cleveland, Ohio 44108

Amax Nickel Inc. *Nickel*
1 Greenwich Plaza
Greenwich, Conn. 06831

Beardsley & Piper *Mullers*
2541 N. Keeler Ave.
Chicago, Ill. 60639

Herman Corp. *Molding machines*
West New Castle St.
Zelienople, Pa. 16063

A. P. Green Refractories
Green Blvd.
Mexico, Mo. 65265

Combustion Engineering Inc. *Ladles etc.*
7887 Hub Parkway
Cleveland, Ohio 44125

Stauffer Chemical Co. *Ethyl silicate*
Speciality Chem. Div.
Westport, Conn. 06880

International Nickel Co. Inc. *Nickel*
New York, N. Y. 10004

Royer Foundry & Machine Co. *Sand Cond.*
159 Pringle St.
Kingston, Pa. 18704

Wheelabrator Frye Inc.
505 S. Byrkit
Michawaka, Ind. 46544

National Engineering Co. *Mullers*
20 North Wacker Dr.
Chicago, Ill. 60606

Black & Decker
Air Tool Division *Air Tools*
6225 Cochran Rd.
Solon, Ohio 44139

Great Western Mfg. Co. *Miscellaneous*
Leavenworth, Kansas 66048

Hauck Mfg. Co. *Miscellaneous*
P. O. Box 90
Lebanon, Penn. 17042

Joseph Dixon *Crucibles*
Div. 48-C
Wayne & Monmouth St.
Jersey City, N. J. 07303

Wedron Silica
Pebble Beach Corp.
400 West Higgins Rd.
Park Ridge, Ill. 60068

EGC Enterprises Inc. *Miscellaneous*
7315 Industrial Parkway
Mentor, Ohio 44060

MC-Englevan Heat Treating & Mfg. Co. *Metal melting furnaces*
Box 31 708-T Griggs St.
Danville, Ill. 61832

The Adams Co. *Flasks*
Philip & Adams St.
Dubuque, Iowa 52001

Fremont Flask Co. *Flasks*
1420 Wolfe Ave.
Fremont, Ohio 43420

Wire Tex Mfg. Co. *Hand riddles*
Mason & Picree Sts.
Bridgeport, Conn. 06605

Hines Flask Co. *Flasks*
3431 W. 140th St.
Cleveland, Ohio 44111

Smith & Richardson Mfg. Co. *Chaplets*
705 May St.
Geneva, Ill. 60134

Crescent Chaplets Inc.
5766 Trumbull Ave.
Detroit, Mich. 48202

Index